Use R!

Advisors:
Robert Gentleman • Kurt Hornik • Giovanni Parmigiani

Use R!

Series Editors: Robert Gentleman, Kurt Hornik, and Giovanni Parmigiani

Alain F. Zuur · Elena N. Ieno ·
Erik H.W.G. Meesters

A Beginner's Guide to R

 Springer

Alain F. Zuur
Highland Statistics Ltd.
6 Laverock Road
Newburgh
United Kingdom AB41 6FN
highstat@highstat.com

Elena N. Ieno
Highland Statistics Ltd.
6 Laverock Road
Newburgh
United Kingdom AB41 6FN
bio@highstat.com

Erik H.W.G. Meesters
IMARES, Institute for Marine
    Resources & Ecosystem Studies
1797 SH 't Horntje
The Netherlands
erik.meesters@wur.nl

ISBN 978-0-387-93836-3   e-ISBN 978-0-387-93837-0
DOI 10.1007/978-0-387-93837-0
Springer Dordrecht Heidelberg London New York

Library of Congress Control Number: 2009929643

Printed on acid-free paper

Springer is part of Springer Science + Business Media (www.springer.com)

To my future niece (who will undoubtedly cost me a lot of money)

Alain F. Zuur

To Juan Carlos and Norma

Elena N. Ieno

For Leontine and Ava, Rick, and Merel

Erik H.W.G. Meesters

Preface

The Absolute R Beginner

For whom was this book written?

Since 2000, we have taught statistics to over 5000 life scientists. This sounds a lot, and indeed it is, but with some classes of 200 undergraduate students, numbers accumulate rapidly (although some courses have involved as few as 6 students). Most of our teaching has been done in Europe, but we have also conducted courses in South America, Central America, the Middle East, and New Zealand. Of course teaching at universities and research organisations means that our students may be from almost anywhere in the world. Participants have included undergraduates, but most have been MSc students, postgraduate students, post-docs, or senior scientists, along with some consultants and nonacademics.

This experience has given us an informed awareness of the typical life scientist's knowledge of statistics. The word "typical" may be misleading, as those scientists enrolling in a statistics course are likely to be those who are unfamiliar with the topic or have become rusty. In general, we have worked with people who, at some stage in their education or career, have completed a statistics course covering such topics as mean, variance, t-test, Chi-square test, and hypothesis testing, and perhaps including half an hour devoted to linear regression.

There are many books available on doing statistics with R. But this book does not deal with statistics, as, in our experience, teaching statistics and R at the same time means two steep learning curves, one for the statistical methodology and one for the R code. This is more than many students are prepared to undertake. This book is intended for people seeking an elementary introduction to R. Obviously, the term "elementary" is vague; elementary in one person's view may be advanced in another's.

R contains a high "you need to know what you are doing" content, and its application requires a considerable amount of logical thinking. As statisticians, it is easy to sit in an ivory tower and expect the life scientist to knock on our door and ask to learn our language. This book aims to make that language as simple as possible. If the phrase "absolute beginner" offends, we apologize, but it answers the question: For whom is this book intended?

All authors of this book are Windows users and have limited experience with Linux and with Mac OS. R is also available for computers with these operating systems, and all the R code we present should run properly on them. However, there may be small differences with saving graphs. Non-Windows users will also need to find an alternative to the text editor Tinn-R (Chapter 1 discusses where you can find information on this).

Datasets used in This book

This book uses mainly life science data. Nevertheless, whatever your area of study and whatever your data, the procedures presented will apply. Scientists in all fields need to import data, massage data, make graphs, and, finally, perform analyses. The R commands will be very similar in every case. A 200-page book does not offer a great deal of scope for presenting a variety of dataset types, and, in our experience, widely divergent examples confuse the reader. The optimal approach may be to use a single dataset to demonstrate all techniques, but this does not make many people happy. Therefore, we have used ecological datasets (e.g., involving plants, marine benthos, fish, birds) and epidemiological datasets.

All datasets used in this book are downloadable from www.highstat.com.

Newburgh   Alain F. Zuur
Newburgh   Elena N. Ieno
Den Burg   Erik H.W.G. Meesters

Acknowledgements

We thank Chris Elphick for the sparrow data; Graham Pierce for the squid data; Monty Priede for the ISIT data; Richard Loyn for the Australian bird data; Gerard Janssen for the benthic data; Pam Sikkink for the grassland data; Alexandre Roulin for the barn owl data; Michael Reed and Chris Elphick for the Hawaiian bird data; Robert Cruikshanks, Mary Kelly-Quinn, and John O'Halloran for the Irish river data; Joaquín Vicente and Christian Gortázar for the wild boar and deer data; Ken Mackenzie for the cod data; Sonia Mendes for the whale data; Max Latuhihin and Hanneke Baretta-Bekker for the Dutch salinity and temperature data; and António Mira and Filipe Carvalho for the roadkill data. The full references are given in the text.

This is our third book with Springer, and we thank John Kimmel for giving us the opportunity to write it. We also thank all course participants who commented on the material.

We thank Anatoly Saveliev and Gema Hernádez-Milian for commenting on earlier drafts and Kathleen Hills (The Lucidus Consultancy) for editing the text.

Contents

Chapter 1
Introduction

We begin with a discussion of obtaining and installing R and provide an overview of its uses and general information on getting started. In Section 1.6 we discuss the use of text editors for the code and provide recommendations for the general working style. In Section 1.7 we focus on obtaining assistance using help files and news groups. Installing R and loading packages is discussed in Section 1.8, and an historical overview and discussion of the literature are presented in Section 1.10. In Section 1.11, we provide some general recommendations for reading this book and how to use it if you are an instructor, and finally, in the last section, we summarise the R functions introduced in this chapter.

1.1 What Is R?

It is a simple question, but not so easily answered. In its broadest definition, R is a computer language that allows the user to program algorithms and use tools that have been programmed by others. This vague description applies to many computing languages. It may be more helpful to say what R can do. During our R courses, we tell the students, "R can do anything you can imagine," and this is hardly an overstatement. With R you can write functions, do calculations, apply most available statistical techniques, create simple or complicated graphs, and even write your own library functions. A large user group supports it. Many research institutes, companies, and universities have migrated to R. In the past five years, many books have been published containing references to R and calculations using R functions. A nontrivial point is that R is available free of charge.

Then why isn't everyone using it? This is an easier question to answer. R has a steep learning curve! Its use requires programming, and, although various graphical user interfaces exist, none are comprehensive enough to completely avoid programming. However, once you have mastered R's basic steps, you are unlikely to use any other similar software package.

The programming used in R is similar across methods. Therefore, once you have learned to apply, for example, linear regression, modifying the code so that it does generalised linear modelling, or generalised additive modelling, requires only the modification of a few options or small changes in the formula. In

A.F. Zuur et al., *A Beginner's Guide to R*, Use R,
DOI 10.1007/978-0-387-93837-0_1, © Springer Science+Business Media, LLC 2009

addition, R has excellent statistical facilities. Nearly everything you may need in terms of statistics has already been programmed and made available in R (either as part of the main package or as a user-contributed package).

There are many books that discuss R in conjunction with statistics (Dalgaard, 2002; Crawley, 2002, 2005; Venables and Ripley, 2002; among others. See Section 1.10 for a comprehensive list of R books). This book is not one of them. Learning R and statistics simultaneously means a double learning curve. Based on our experience, that is something for which not many people are prepared. On those occasions that we have taught R and statistics together, we found the majority of students to be more concerned with successfully running the R code than with the statistical aspects of their project. Therefore, this book provides basic instruction in R, and does not deal with statistics. However, if you wish to learn both R and statistics, this book provides a basic knowledge of R that will aid in mastering the statistical tools available in the program.

1.2 Downloading and Installing R

We now discuss acquiring and installing R. If you already have R on your computer, you can skip this section.

The starting point is the R website at www.r-project.org. The homepage (Fig. 1.1) shows several nice graphs as an appetiser, but the important feature is

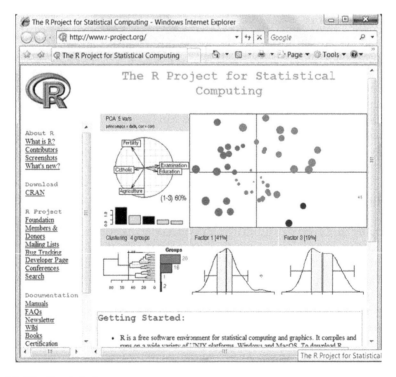

Fig. 1.1 The R website homepage

the **CRAN** link under **Download**. This cryptic notation stands for Comprehensive R Archive Network, and it allows selection of a regional computer network from which you can download R. There is a great deal of other relevant material on this site, but, for the moment, we only discuss how to obtain the R installation file and save it on your computer.

If you click on the CRAN link, you will be shown a list of network servers all over the planet. Our nearest server is in Bristol, England. Selecting the Bristol server (or any of the others) gives the webpage shown in Fig. 1.2. Clicking the Linux, MacOS X, or Windows link produces the window (Fig. 1.3) that allows us to choose between the base installation file and contributed packages. We discuss packages later. For the moment, click on the link labelled **base**.

Clicking **base** produces the window (Fig. 1.4) from which we can download R. Select the setup program **R-2.7.1-win32.exe** and download it to your computer. Note that the size of this file is 25–30 Mb, not something you want to download over a telephone line. Newer versions of R will have a different designation and are likely to be larger.

To install R, click the downloaded **R-2.7.1-win32.exe** file. The simplest procedure is to accept all default settings. Note that, depending on the computer settings, there may be issues with system administration privileges, firewalls, VISTA security settings, and so on. These are all computer- or network-specific problems and are not further discussed here. When you have installed R, you will have a blue desktop icon.

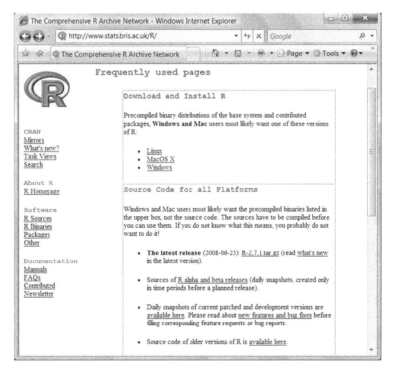

Fig. 1.2 The R local server page. Click the Linux, MacOS X, or Windows link to go to the window in Fig. 1.3

Fig. 1.3 The webpage that allows a choice of downloading R base or contributed packages

To upgrade an installed R program, you need to follow the downloading process described above. It is not a problem to have multiple R versions on your computer; they will be located in the same R directory with different subdirectories and will not influence one another. If you upgrade from an older R version, it is worthwhile to read the CHANGES files. (Some of the information in the CHANGES file may look intimidating, so do not pay much attention to it if you are a novice user.)

1.3 An Initial Impression

We now discuss opening the R program and performing some simple tasks. Startup of R depends upon how it is installed. If you have downloaded it from www.r-project.org and installed it on a standalone computer, R can be started by double-clicking the desktop shortcut icon or by going to **Start- > Program- > R**. On network computers with a preinstalled version, you may need to ask your system administrator where to find the shortcut to R.

The program will open with the window in Fig. 1.5. This is the starting point for all that is to come.

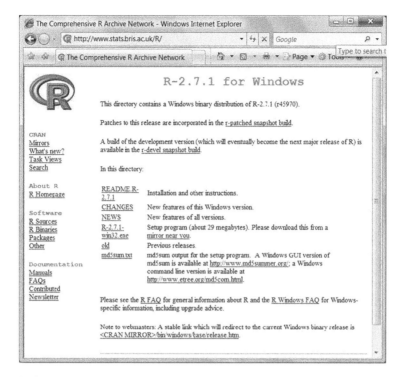

Fig. 1.4 The window that allows you to download the setup file **R-2.7.1-win32.exe**. Note that this is the latest version at the time of writing, and you may see a more recent version

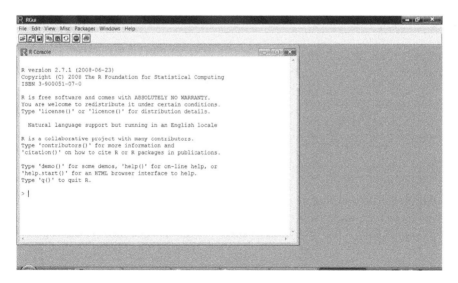

Fig. 1.5 The R startup window. It is also called the console or command window

There are a few things that are immediately noticeable from Fig. 1.5. (1) the R version we use is 2.7.1; (2) there is no nice looking graphical user interface (GUI); (3) it is free software and comes with absolutely no warranty; (4) there is a help menu; and (5) the symbol > and the cursor. As to the first point, it does not matter which version you are running, provided it is not too dated. Hardly any software package comes with a warranty, be it free or commercial. The consequence of the absence of a GUI and of using the help menu is discussed later. Moving on to the last point, type 2 + 2 after the > symbol (which is where the cursor appears):

```
> 2 + 2
```

and click enter. The spacing in your command is not relevant. You could also type 2 + 2, or 2 + 2. We use this simple R command to emphasise that you must type something into the command window to elicit output from R. 2 + 2 will produce:

```
[1] 4
```

The meaning of [1] is discussed in the next chapter, but it is apparent that R can calculate the sum of 2 and 2. The simple example shows how R works; you type something, press enter, and R will carry out your commands. The trick is to type in sensible things. Mistakes can easily be made. For example, suppose you want to calculate the logarithm of 2 with base 10. You may type:

```
> log(2)
```

and receive:

```
[1] 0.6931472
```

but 0.693 is not the correct answer. This is the natural logarithm. You should have used:

```
> log10(2)
```

which will give the correct answer:

```
[1] 0.30103
```

Although the log and log10 command can, and should, be committed to memory, we later show examples of code that is impossible to memorise. Typing mistakes can also cause problems. Typing 2 + 2w will produce the message

```
> 2 + 2w
Error: syntax error in "2+2w"
```

R does not know that the key for w is close to 2 (at least for UK keyboards), and that we accidentally hit both keys at the same time.

The process of entering code is fundamentally different from using a GUI in which you select variables from drop-down menus, click or double-click an option and/or press a "go" or "ok" button. The advantages of typing code are that it forces you to think what to type and what it means, and that it gives more flexibility. The major disadvantage is that you need to know what to type.

R has excellent graphing facilities. But again, you cannot select options from a convenient menu, but need to enter the precise code or copy it from a previous project. Discovering how to change, for example, the direction of tick marks, may require searching Internet newsgroups or digging out online manuals.

1.4 Script Code

1.4.1 The Art of Programming

At this stage it is not important that you understand anything of the code below. We suggest that you do not attempt to type it in. We only present it to illustrate that, with some effort, you can produce very nice graphs using R.

```
> setwd("C:/RBook/")
> ISIT<-read.table("ISIT.txt",header=TRUE)
> library(lattice)
> xyplot(Sources~SampleDepth|factor(Station),data=ISIT,
  xlab="Sample Depth",ylab="Sources",
  strip=function(bg='white', ...)
  strip.default(bg='white', ...),
  panel = function(x, y) {
  panel.grid(h=-1, v= 2)
  I1<-order(x)
  llines(x[I1], y[I1],col=1)})
```

All the code from the third line (where the xyplot starts) onward forms a single command, hence we used only one > symbol. Later in this section, we improve the readability of this script code. The resulting graph is presented in Fig. 1.6. It plots the density of deep-sea pelagic bioluminescent organisms versus depth for 19 stations. The data were gathered in 2001 and 2002 during a series of four cruises of the *Royal Research Ship Discovery* in the temperate NE Atlantic west of Ireland (Gillibrand et al., 2006). Generating the graph took considerable effort, but the reward is that this single graph gives all the information and helps determine which statistical methods should be applied in the next step of the data analysis (Zuur et al., 2009).

Fig. 1.6 Deep-sea pelagic bioluminescent organisms versus depth (in metres) for 19 stations. Data were taken from Zuur et al. (2009). It is relatively easy to allow for different ranges along the *y*-axes and *x*-axes. The data were provided by Monty Priede, Oceanlab, University of Aberdeen, Aberdeen, UK

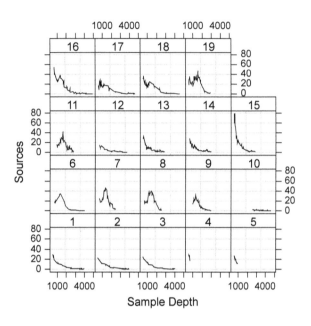

1.4.2 Documenting Script Code

Unless you have an exceptional memory for computing code, blocks of R code, such as those used to create Fig. 1.6, are nearly impossible to remember. It is therefore fundamentally important that you write your code to be as general and simple as possible and document it religiously. Careful documentation will allow you to reproduce the graph (or other analysis) for another dataset in only a matter of minutes, whereas, without a record, you may be alienated from your own code and need to reprogram the entire project. As an example, we have reproduced the code used in the previous section, but have now added comments. Text after the symbol "#" is ignored by R. Although we have not yet discussed R syntax, the code starts to make sense. Again, we suggest that you do not attempt to type in the code at this stage.

```
> setwd("C:/RBook/")
> ISIT<-read.table("ISIT.txt",header=TRUE)
#Start the actual plotting
#Plot Sources as a function of SampleDepth, and use a
#panel for each station.
#Use the colour black (col=1), and specify x and y
#labels (xlab and ylab). Use white background in the
#boxes that contain the labels for station
```

```
>xyplot(Sources~SampleDepth|factor(Station),
   data = ISIT,xlab="Sample Depth",ylab="Sources",
   strip=function(bg='white', ...)
   strip.default(bg='white', ...),
   panel = function(x,y) {
       #Add grid lines
       #Avoid spaghetti plots
       #plot the data as lines (in the colour black)
         panel.grid(h=-1,v= 2)
         I1<-order(x)
         llines(x[I1],y[I1],col=1)})
```

Although it is still difficult to understand what the code is doing, we can at least detect some structure in it. You may have noticed that we use spaces to indicate which pieces of code belong together. This is a common programming style and is essential for understanding your code. If you do not understand code that you have programmed in the past, do not expect that others will! Another way to improve readability of R code is to add spaces around commands, variables, commas, and so on. Compare the code below and above, and judge for yourself what looks easier. We prefer the code below (again, do not attempt to type the code).

```
> setwd("C:/RBook/")
> ISIT <- read.table("ISIT.txt", header = TRUE)
> library(lattice) #Load the lattice package

#Start the actual plotting
#Plot Sources as a function of SampleDepth, and use a
#panel for each station.
#Use the colour black (col=1), and specify x and y
#labels (xlab and ylab). Use white background in the
#boxes that contain the labels for station
> xyplot(Sources ~ SampleDepth | factor(Station),
          data = ISIT,
          xlab = "Sample Depth", ylab = "Sources",
          strip = function(bg = 'white', ...)
          strip.default(bg = 'white', ...),
          panel = function(x, y) {
            #Add grid lines
            #Avoid spaghetti plots
            #plot the data as lines (in the colour black)
          panel.grid(h = -1, v = 2)
          I1 <- order(x)
          llines(x[I1], y[I1], col = 1)})
```

We later discuss further steps that can be taken to improve the readability of this particular piece of code.

1.5 Graphing Facilities in R

One of the most important steps in data analysis is visualising the data, which requires software with good plotting facilities. The graph in Fig. 1.7, showing the laying dates of the Emperor Penguin (*Aptenodytes forsteri*), was created in R with five lines of code. Barbraud and Weimerskirch (2006) and Zuur et al. (2009) looked at the relationship of arrival and laying dates of several bird species to climatic variables, measured near the Dumont d'Urville research station in Terre Adélie, East Antarctica.

Fig. 1.7 Laying dates of Emperor Penguins in Terre Adélie, East Antarctica. To create the background image, the original jpeg image was reduced in size and exported to portable pixelmap (ppm) from a graphics package. The R package `pixmap` was used to import the background image into R, the `plot` command was applied to produce the plot and the `addlogo` command overlaid the ppm file. The photograph was provided by Christoph Barbraud

It is possible to have a small penguin image in a corner of the graph, or it can also be stretched so that it covers the entire plotting region.

Whilst it is an attractive graph, its creation took three hours, even using sample code from Murrell (2006). Additionally, it was necessary to reduce the resolution and size of the photo, as initial attempts caused serious memory problems, despite using a recent model computer.

Hence, not all things in R are easy. The authors of this book have often found themselves searching the R newsgroup to find answers to relatively simple questions. When asked by an editor to alter line thickness in a complicated multipanel graph, it took a full day. However, whereas the graph with the penguins could have been made with any decent graphics package, or even in Microsoft Word, we show graphs that cannot be easily made with any other program.

Figure 1.8 shows the nightmare of many statisticians, the Excel menu for pie charts. Producing a scientific paper, thesis, or report in which the only graphs are pie charts or three-dimensional bar plots is seen by many experts as a sign of incompetence. We do not wish to join the discussion of whether a pie chart is a good or bad tool. Google "pie chart bad" to see the endless list of websites expressing opinions on this. We do want to stress that R's graphing tools are a considerable improvement over those in Excel. However, if the choice is between the menu-driven style in Fig. 1.8 and the complicated looking code given in Section 1.3, the temptation to use Excel is strong.

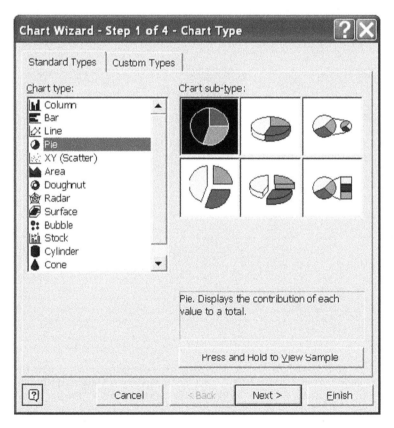

Fig. 1.8 The pie chart menu in Excel

1.6 Editors

As explained above, the process of running R code requires the user to type the code and click enter. Typing the code into a special text editor for copying and pasting into R is strongly recommended. This allows the user to easily save code, document it, and rerun it at a later stage. The question is which text editor to use. Our experience is with Windows operating systems, and we are unable to recommend editors for Mac, UNIX, or LINUX. A detailed description of a large number of editors is given at http://www.sciviews.org/_rgui/projects/Editors.html. This page contains some information on Mac, UNIX, and LINUX editors.

For Windows operating systems, we strongly advise against using Microsoft Word. Word automatically wraps text over multiple lines and adds capitals to words at the beginning of the line. Both will cause error messages in R. R's own text editor (click **File- > New script** as shown in Fig. 1.5) and Notepad are alternatives, although neither have the bells and whistles available in R-specific text editors such as Tinn-R (http://www.sciviews.org/Tinn-R/) and RWindEdt (this is an R package).

R is case sensitive, and programming requires the use of curly brackets {}, round brackets (), and square brackets []. It is important that an opening bracket

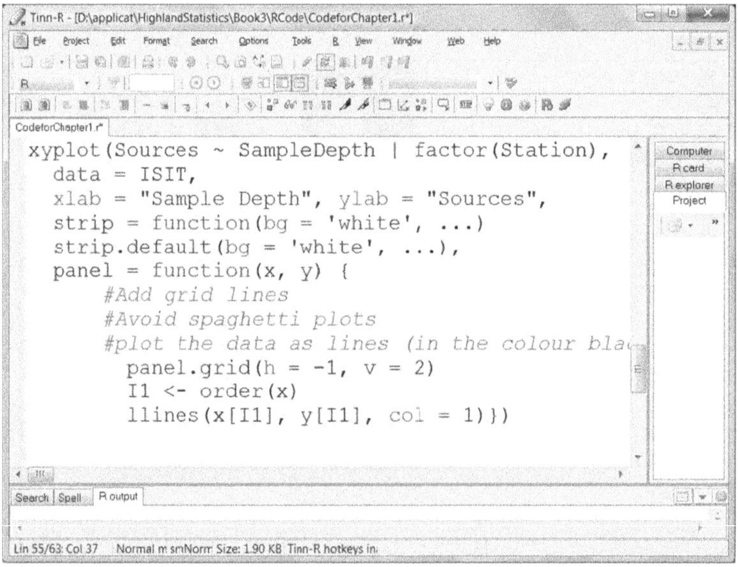

Fig. 1.9 The Tinn-R text editor. Each bracket style has a distinctive colour. Under **Options- > Main- > Editor**, the font size can be increased. Under **Options- > Main- > Application- > R**, you can specify the path for R. Select the **Rgui.exe** file in the directory **C:\Program Files\R\R-2.7.1\bin** (assuming default installation settings). Adjust the R directory if you use a different R version. This option allows sending blocks of code directly to R by highlighting code and clicking one of the icons above the file name

{ is matched by a closing bracket } and that it is used in the correct position for the task. Some of the errors made by an R novice are related to omitting a bracket or using the wrong type of bracket. Tinn-R and RWinEdt use colours to indicate matching brackets, and this is an extremely useful tool. They also use different colours to identify functions from other code, helping to highlight typing mistakes.

Tinn-R is available free, whereas RWinEdt is shareware and requires a small payment after a period of time. Both programs allow highlighting text in the editor and clicking a button to send the code directly to R, where it is executed. This bypasses copying and pasting, although the option may not work on some network systems. We refer to the online manuals of Tinn-R and RWinEdt for their use with R.

A snapshot of Tinn-R, our preferred editor, is shown in Fig. 1.9. To re-emphasise, write your R code in an editor such as Tinn-R, even if it is only a few commands, before copying and pasting (or sending it directly) to R.

1.7 Help Files and Newsgroups

When working in R, you will have multiple options for nearly every task, and, because there is no single source that describes all the possibilities, it is essential that you know where to look for help. Suppose you wish to learn to make a boxplot. Your best friend in R is the question mark. Type:

```
> ?boxplot
```

and hit the enter key. Alternatively, you can also use:

```
> help(boxplot)
```

A help window opens, showing a document with the headings *Description, Usage, Arguments, Details, Values, References, See also,* and *Examples.* These help files are not "guides for dummies" and may look intimidating. We recommend that you read the description, quickly browse the usage section (marvelling at the undecipherable options), and proceed to the examples to get an idea of R's boxplot capabilities. Copy some of the sample code and paste it into R.

The following lines of code from the example in the help file,

```
> boxplot(count ~ spray, data = InsectSprays,
        col = "lightgray")
```

produce the boxplot in Fig. 1.10. The syntax, *count ~ spray*, ensures that one boxplot per level of insect sprays is generated. Information on the insect spray data can be obtained by typing:

```
> ?InsectSprays
```

Fig. 1.10 Boxplot obtained
by copying and pasting code
from the boxplot help file
into R. To see the data on
which the graph is based,
type: ? InsectSprays

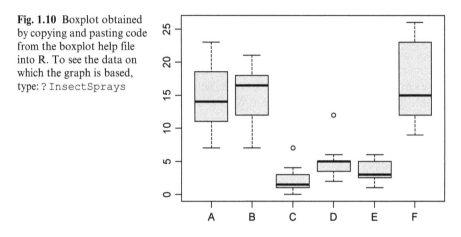

It is important to copy entire pieces of code and not a segment that contains only part of the code. With long pieces of code, it can be difficult to identify beginning and end points, and sometimes guesswork is needed to determine where a particular command ends. For example, if you only copy and paste the text

```
> boxplot(count ~ spray, data = InsectSprays,
```

you will see a "+" symbol (Fig. 1.11), indicating that R expects more code. Either paste in the remainder of the code, or press *escape* to cancel the action and proceed to copy and paste in the entire command.

Nearly all help files have a structure similar to the help file of the *boxplot* function.

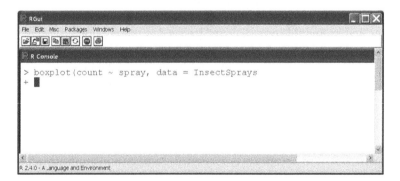

Fig. 1.11 R is waiting for more code, as an incomplete command has been typed. Either add the remaining code or press "escape" to abort the boxplot command

If you cannot find the answer in a help file, click **Help- > Html help** in the menu (Fig. 1.5). The window in Fig. 1.12 will appear (provided your pop-up blocker is switched off), and the links in the browser will provide a wealth of information. The **Search Engine & Keywords** link allows you to search for functions, commands, and keywords.

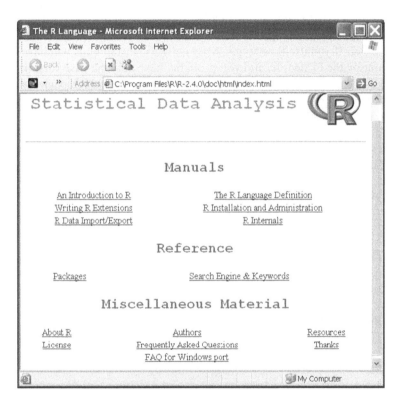

Fig. 1.12 The window that is obtained by clicking **Help- > Html help** from the help menu in R. *Search Engine & Keywords* allows searching for functions, commands, and keywords. You will need to switch off any pop-up blockers

If the help files haven't provided the answer to your question(s), it is time for a search on the R newsgroup. It is likely that others have discussed your question in the past. The R newsgroup can be found by going to www.r-project.org. Click *Mailing Lists,* go to the R-help section, and click *web-interface.* To access the hundreds of thousands of postings go to one of the searchable archives. It is now a matter of using relevant keywords to find similar problems.

If you still cannot find the answer to your question, then as a last resort you can try posting a message to the newsgroup. First read the posting guidelines, or you may be reminded that you should have done so, especially if your question turns out to have been discussed before, or is answered in the help files.

1.8 Packages

R comes with a series of default packages. A package is a collection of pre-viously programmed functions, often including functions for specific tasks. It is tempting to call this a library, but the R community refers to it as a package.

There are two types of packages: those that come with the base installation of R and packages that you must manually download and install. With the base installation we mean the big executable file that you downloaded and installed in Section 1.2. If you use R on a university network, installing R will have been carried out by the IT people, and you probably have only the base version. The base version contains the most common packages. To see which packages you have, click **Packages - > Load package** (Fig. 1.5).

There are literally hundreds of user-contributed packages that are not part of the base installation, most of which are available on the R website. Many packages are available that will allow you to execute the same statistical calculations as commercial packages. For example, the multivariate vegan package can execute methods that are possible using commercial packages such as PRIMER, PCORD, CANOCO, and Brodgar.

1.8.1 Packages Included with the Base Installation

Loading a package that came with the base installation may be accomplished either by a mouse click or by entering a specific command.

You can click **Packages- > Load package** (Fig. 1.5), select a package, and click *ok*. Those who hate clicking (as we do), may find it more efficient to use the library command. For instance, to load the MASS package, type the command:

```
> library(MASS)
```

and press enter. You now have access to all functions in the MASS package. So what next? You could read a book, such as that by Venables and Ripley (2002), to learn all that you can do with this package. More often the process is reversed. Suppose you have a dataset to which you want to apply generalised linear mixed modelling (GLMM).[1] Consulting Venables and Ripley (2002) will show that you can do this with the function **glmmPQL** in the MASS package (other options exist). Hence, you load MASS with the library command as explained above and type ? glmmPQL to receive instructions in applying GLMM.

---

[1] A GLMM is an advanced linear regression model. Instead of the Normal distribution, other types of distributions can be used, for example, the Poisson or negative binomial distribution for count data and the binomial distribution for binary data.

1.8.2   Packages Not Included with the Base Installation

Sometimes the process of loading a package is slightly more complicated. For example, suppose you see a paper in which data are plotted versus their spatial locations (latitude and longitude), and the size of the dots is proportional to the data values. The text states that the graph was made with the bubble function from the gstat package. If you click **Packages- > Load package** (as shown in Fig. 1.5), you will not see gstat. If a package does not appear in the list, it has not been installed. Hence this method can also be used to determine whether a package is part of the base installation. To obtain and install gstat, or any other available package, you can download the zipped package from the R website and tell R to install it, or you can install it from within R. We discuss both options. In addition there is a third option, which is described in the help file of the function install.packages

Note that the process of installing a package need only be done once.

Option 1. Manual Download and Installation

On your Internet browser, go to the R website (www.r-project.org), click CRAN, select a server, and click **Packages** under the **Software** heading. You are presented with a list of user-contributed packages. Select gstat (which is a long way down). You can now download the zipped package (for Windows operating systems this is the file called Windows binary) and a manual. Once you have downloaded the file to your hard disk, go to R and click **Packages- > Install packages** from **local zip file**. Select the file that you just downloaded.

The websites for packages generally have a manual in PDF format which may provide additional useful information. A potential problem with manual downloads is that sometimes a package is dependent upon other packages that are also not included in the base installation, and you need to download those as well. Any dependencies on other packages are mentioned on the website, but it can be like a family tree; these secondary packages may be dependent on yet other packages.

The following method installs any dependent packages automatically.

Option 2. Download and Install a Package from Within R

As shown in Fig. 1.5, click **Packages- > set the CRAN mirror** and **select a server** (e.g., Bristol, UK). Now go back to **Packages** and click **Install package(s)** which will present a list of packages from which you can select gstat. You only need to execute this process once, unless you update R to a newer version. (Updates appear on a regular basis, but there is no need to update R as soon as there is a new version. Updating once or twice per year is enough.)

Note that there may be installation problems on networked computers, and when using Windows VISTA, related to your firewall or other security settings. These are computer-specific problems, and are not discussed here.

1.8.2.1 Loading the Package

There is a difference between installing and loading. *Install* denotes adding the package to the base version of R. *Load* means that we can access all the functions in the package, and are ready to use it. You cannot load a package if it is not installed. To load the `gstat` package we can use one of the two methods described in Section 1.8.1. Once it has been loaded, **?bubble** will give instructions for using the function.

We have summarised the process of installing and loading packages in Fig. 1.13.

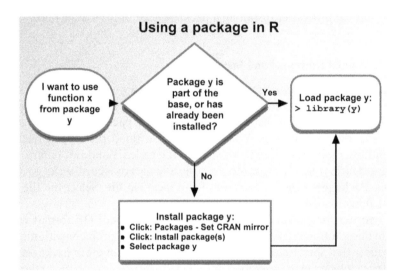

Fig. 1.13 Overview of the process of installing and loading packages in R. If a package is part of the base installation, or has previously been installed, use the `library` function. If a package's operation depends upon other packages, they will be automatically loaded, provided they have been installed. If not, they can be manually installed. Once a package has been installed, you do not have to install it again

1.8.2.2 How Good Is a Package?

During courses, participants sometimes ask about the quality of these user-contributed packages. Some of the packages contain hundreds of functions written by leading scientists in their field, who have often written a book in which the methods are described. Other packages contain only a few functions that may have been used in a published paper. Hence, you have packages from a range of contributors from the enthusiastic PhD student to the professor who has published ten books. There is no way to say which is better. Check how often a package has been updated, and enter it into the R newsgroup to see other's experiences.

1.9 General Issues in R

In this section, we discuss various issues in working with R, as well as methods of making the process simpler.

If you are an instructor who gives presentations using R, or if you have difficulties reading small-sized text, the ability to adjust font size is essential. This can be done in R by clicking **Edit- < GUI preferences**.

First-time users may be confused by the behaviour of the console once a graph has been made. For an example, see Fig. 1.14. Note that the graphic device is active. If you attempt to copy and paste code into R, there will be no response. You need to make the R console window (on the left) active before you can paste R code. If the R console window is maximised when pasting code, the graphic device (behind the R console window) will not be visible. Either change the size of the console window, or use the CRTL/TAB keys to alternate between windows.

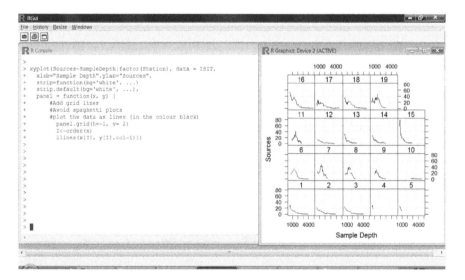

Fig. 1.14 R after making a graph. To run new commands, you must first click on the console

To save a graph, click to make it active and right-click the mouse. You can then copy it as a metafile directly into another program such as Microsoft Word. Later, we discuss commands to save graphics to files.

A common mistake that many people make when using Tinn-R (or any other text editor) is that they do not copy the "hidden enter" notation on the last line

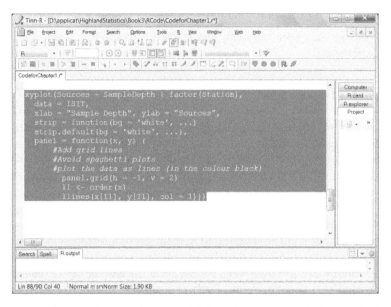

Fig. 1.15 Our Tinn-R code. Note that we copied the code up to, and including, the final round bracket. We should have dragged the mouse one line lower to include the hidden *enter* that will execute the `xyplot` command

of code. To show what we mean by this, see Figs. 1.15 and 1.16. In the first figure, we copied the R code for the `xyplot` command previously entered into Tinn-R. Note that we stopped selecting immediately after the final round bracket. Pasting this segment of code into R produces Fig. 1.16. R is now waiting for us to press enter, which will make the graph appear. This situation can cause panic as R seems to do nothing even though the code is correct and was completely copied into R—with the exception of the *enter* command on the final line of the code. The solution is simple: press enter, and, next time, highlight an extra line beneath the final round bracket before copying.

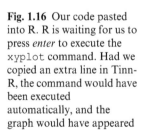

Fig. 1.16 Our code pasted into R. R is waiting for us to press *enter* to execute the `xyplot` command. Had we copied an extra line in Tinn-R, the command would have been executed automatically, and the graph would have appeared

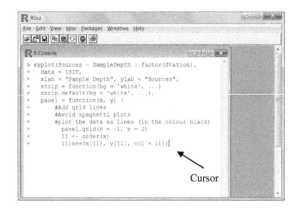

1.9.1 Quitting R and Setting the Working Directory

Another useful command is:

```
> q()
```

It exits R. Before it does so, it will ask whether it should save the workspace. If you decide to save it, we strongly advise that you do not save it in its default directory. Doing so will cause R to load all your results automatically when it is restarted. To avoid R asking whether it should save your data, use:

```
> q(save = "no")
```

R will then quit without saving. To change the default working directory use:

```
> setwd(file = "C:\\AnyDirectory\\")
```

This command only works if the directory `AnyDirectory` already exists; choose a sensible name (ours isn't). Note that you must use two backward slashes on Windows operating systems. The alternative is to use:

```
> setwd(file = "C:/AnyDirectory/")
```

Use simple names in the directory structure. Avoid directory names that contain symbols such as *, &, ○, $, £, ", and so on. R also does not accept alphabetic symbols with diacritical marks, ä, í, á, ö, è, é, and so on.

Our recommendation is that, rather than saving your workspace, you save your R code in your text editor. Next time, open your well-documented saved file, copy the code, and paste it into R. Your results and graphs will reappear. Saving your workspace only serves to clutter your hard disk with more files, and also in a week's time you may not remember how you obtained all the variables, matrices, and so on. Retrieving this information from your R code is much easier. The only exception is if your calculations take a long time to complete. If this is the case, it's advisable to save the workspace somewhere in your working directory. To save a workspace, click **File- < Save Workspace**. To load an existing workspace, use **File- < Load Workspace**.

If you want to begin a new analysis on a different dataset, it may be useful to remove all variables. One option is to quit R and restart it. Alternatively, click **Misc- < Remove all objects**. This will execute the command

```
> rm(list = ls(all = TRUE))
```

Other useful options can be found under **Edit**. For example, you can click **Select all** and copy every command and output to Microsoft Word.

1.10 A History and a Literature Overview

1.10.1 A Short Historical Overview of R

If you are ready to begin working with R, a history lesson is the last thing you want. However, we can guarantee that at some stage someone is going to ask you why the package is called R. To provide you with an impressive response, we spend a few words on how, why, and when the package was developed, as well as by whom. After all, a bit of historical knowledge does no harm!

R is based on the computer language S, which was developed by John Chambers and others at Bell Laboratories in 1976. In the early 1990s, Ross Ihaka and Robert Gentleman (University of Auckland, in Auckland, New Zealand) experimented with the language and called their product R. Note that both their first names begin with the letter R. Also when modifying a language called S, it is tempting to call it T or R.

Since 1997, R has been developed by the R Development Core Team. A list of team members can be found at The R FAQ (Hornik, 2008; http://CRAN.R-project.org/doc/FAQ/R-FAQ.html).

The Wikipedia website gives a nice overview of R milestones. In 2000, version 1.0.0 was released, and since then various extensions have been made available. This book was written using version 2.7, which was released in April 2008.

1.10.2 Books on R and Books Using R

The problem with providing an overview of books using R is that there is a good chance of omitting some books, or writing a purely subjective overview. There is also a time aspect involved; by the time you read this, many new books on R will have appeared. Hence, we limit our discussion to books that we have found useful.

Although there are surprisingly few books *on* R; many *use* R to do statistics. We do not make a distinction between these.

Our number one is *Statistical Models in S*, by Chambers and Hastie (1992), informally called the white book as it has a white cover. It does not deal directly with R, but rather with the language on which R is based. However, there is little practical difference. This book gives a detailed explanation of S and how to apply a large number of statistical techniques in S. It also contains some statistical theory.

Our second most used book is *Modern Applied Statistics with S, 4th ed.*, by Venables and Ripley (2002), closely followed by *Introductory Statistics with R* from Dalgaard (2002). At the time of this writing, the second edition of Dalgaard is in press. Both books are "must-haves" for the R user.

There are also books describing general statistical methodology that use R in the implementation. Some of those on our shelves, along with our assessment, are:

- *The R book*, by Crawley (2007). This is a hefty book which quickly intro-duces a wide variety of statistical methods and demonstrates how they can be applied in R. A disadvantage is that once you start using a particular method, you will need to obtain further literature to dig deeper into the underlying statistical methodology.

- *Statistics. An Introduction Using R*, by Crawley (2005).
- *A Handbook of Statistical Analysis Using R*, by Everitt and Hothorn (2006).
- *Linear Models with R*, by Faraway (2005). We highly recommend this book, as well as its sequel, *Extending the Linear Model with R,* from the same author.
- *Data Analysis and Graphics Using R: An Example-Based Approach*, by Maindonald and Braun (2003). This book has a strong regression and generalised linear modelling component and also some general text on R.
- *An R and S-PLUS Companion to Multivariate Analysis*, by Everitt (2005). This book deals with classical multivariate analysis techniques, such as factor analysis, multidimensional scaling, and principal component analysis, and also contains a mixed effects modelling chapter.
- *Using R for Introductory Statistics* by Verzani (2005). The title describes the content; it is useful for an undergraduate statistics course.
- *R Graphics* by Murrell (2006). A "must-have" if you want to delve more deeply into R graphics.

There are also a large number of more specialised books that use R, for example:

- *Time Series Analysis and Its Application. With R Examples — Second Edition*, by Shumway and Stoffer. This is a good time series book.
- *Data Analysis Using Regression and Multilevel/Hierarchical Models*, by Gelman and Hill. A book on mixed effects models for social science using R code and R output.
- In mixed effects models, the "must-buy" and "must-cite" book is *Mixed Effects Models in S and S-Plus*, from Pinheiro and Bates (2000).
- On the same theme, the "must-buy" and "must-cite" book for generalised additive modelling is *Generalized Additive Models: An Introduction with R*, by Wood (2006).
- The latter two books are not easy to read for the less mathematically oriented reader, and an alternative is *Mixed Effects Models and Extensions in Ecology with R*, by Zuur et al. (2009). Because its first two authors are also authors of the book that you are currently reading, it is a "must buy immediately" and "must read from A to Z" book!
- Another easy-to-read book on generalised additive modelling with R is *Semi-Parametric Regression for the Social Sciences*, by Keele (2008).
- If you work with genomics and molecular data, *Bioinformatics and Computational Biology Solutions Using R and Bioconductor*, by Gentleman et al. (2005) is a good first step.
- We also highly recommend *An R and S-Plus Companion to Applied Regression*, from Fox (2002).
- At the introductory level, you may want to consider *A First Course in Statistical Programming with R*, by Braun and Murdoch (2007).

- Because we are addicted to the `lattice` package with its beautiful multi-panel figures (see Chapter 8), we highly recommend *Lattice. Multivariate Data Visualization with R* written by Sarkar (2008). This book has not left our desk since it arrived.

1.10.2.1 The Use R! Series

This book is a part of the Springer series "Use R!," which at the time of writing comprises at least 15 books, each describing a particular statistical method and its use in R, with more books being in press.

If you are lucky, your statistical problem is discussed in one of the books in this series. For example, if you work with morphometric data, you should definitely have a look at *Morphometrics with R*, from Claude (2008). For spatial data try *Applied Spatial Data Analysis with R,* by Bivand et al. (2008), and for wavelet analysis, see *Wavelet Methods in Statistics with R*, by Nason (2008). Another useful volume in this series is *Data Manipulation with R*, from Spector (2008); no more tedious Excel or Access data preparation with this book! For further suggestions we recommend that you consult http://www.springer.com/series/6991 for an updated list.

We have undoubtedly omitted books, and in so doing may have upset readers and authors, but this is what we have on our shelves at the time of writing. A more comprehensive list can be found at: http://www.r-project.org/doc/bib/R-publications.html.

1.11 Using This Book

Before deciding which chapters you should focus on and which you can skip upon first reading, think about the question, "Why would I use R?" We have heard a wide variety of answers to this question, such as:

1. My colleagues are using it.
2. I am interested in it.
3. I need to apply statistical techniques that are only available in R.
4. It is free.
5. It has fantastic graphing facilities.
6. It is the only statistics package installed on the network.
7. I am doing this as part of an education programme (e.g., BSc, MSc, PhD).
8. I have been told to do this by my supervisor.
9. It is in my job description to learn R.

In our courses, we've had a range of participants from the unmotivated, "I have been told to do it" to the supermotivated, "I am interested." How you can best use this book depends somewhat on your own motivation for learning R. If you are the, "I am interested," person, read this book from A to Z. The following gives general recommendations on consuming the information presented, depending on your own situation.

Some of the sections in this book are marked with an asterisk (*); these are slightly more technical, and you may skip them upon first reading.

1.11.1 If You Are an Instructor

Because the material in this book has been used in our own R and statistics courses, we have seen the reactions of many students exposed to it. Our first recommendation is simple: *Do not go too fast!* You will waste your time, and that of your students, by trying to cover as much material as possible in a one or two-day R course. We have taught statistics (and R) to over 5000 life scientists and found the main element in positive feedback to be ensuring that the participants understand what they have been doing. Most participants begin with a "show me all" mentality, and it is your task to change this to "understand it all."

No one wants to do a five-day R course, and this is not necessary. We recommend three-day courses (where a day is eight hours), with the title "Introduction to R." On the first day, you can cover Chapters 1, 2, and 3, and give plenty of exercises. On the second day, introduce basic plotting tools (Chapter 5), and, depending on aims and interests, you can either continue with making functions (Chapter 6) or advanced plotting tools (Chapters 7 and 8) on day three. Chapter 9 contains common mistakes, and these are relevant for everyone.

If you proceed more rapidly, you are likely to end up with frustrated participants. Our recommendation is not to include statistics in such a three-day course. If you do need to cover statistics, extend the course to five days.

1.11.2 If You Are an Interested Reader with Limited R Experience

We suggest reading Chapters 1, 2, 3, and 5. What comes next depends on your interests. Do you want to write your own functions? Chapter 6 is relevant. Or do you want to make fancy graphs? In that case, continue with Chapters 7 and 8.

1.11.3 If You Are an R Expert

If you have experience in using R, we recommend beginning with Chapters 6, 8, and 9.

1.11.4 If You Are Afraid of R

"My colleague has tried R and it was a nightmare. It is beyond many biologists unless they have a very mathematical leaning!" This was taken verbatim from our email inbox, and is indicative of many comments we hear. R is a language, like Italian, Dutch, Spanish, English, or Chinese. Some people have a natural talent for languages, others struggle, and, for some, learning a language is a nightmare. Using R requires that you learn a language. If you try to proceed too rapidly, use the wrong reading material, or have the wrong teacher, then, yes, mastering R may be challenging.

The term "mathematical" comes in because R is a language where tasks proceed in logical steps. Your work in R must be approached in a structured and organized way. But that is essentially all that is necessary, plus a good book.

However, we also want to be honest. Based on our experience, a small fraction of the "typical" scientists attending our courses are not destined to work with R. We have seen people frustrated after a single day of R programming. We have had people tell us that they will never use R again. Luckily, this is only a very small percentage. If you are one of these, we recommend a graphical user interface driven software package such as SPLUS or SAS. These are rather expensive programs. An alternative is to try one of the graphical user interfaces in R (on the R website, select **Related Projects** from the menu **Misc**, and then click **R GUIs**), but these will not give you the full range of options available in R.

1.12 Citing R and Citing Packages

You have access to a free package that is extremely powerful. In recognition, it is appropriate therefore, to cite R, or any associated package that you use. Once in R, type:

```
> citation()

To cite R in publications use:
R Development Core Team (2008). R: A language and
environment for statistical computing. R Foundation for
Statistical Computing, Vienna, Austria.
ISBN 3-900051-07-0, URL http://www.R-project.org.
...
We have invested a lot of time and effort in creating R,
please cite it when using it for data analysis. See also
'citation("pkgname")' for citing R packages.
```

For citing a package, for example the lattice package, you should type:

```
> citation("lattice")
```

It gives the full details on how to cite this package. In this book, we use various packages; we mention and cite them all below: foreign (R-core members et al., 2008), lattice (Sarkar, 2008), MASS (Venables and Ripley, 2002), nlme (Pinheiro et al., 2008), plotrix (Lemon et al., 2008), RODBC (Lapsley, 2002;

Ripley, 2008), and vegan (Oksanen et al., 2008). The reference for R itself is: R Development Core Team (2008). Note that some references may differ depending on the version of R used.

1.13 Which R Functions Did We Learn?

We conclude each chapter with a section in which we repeat the R functions that were introduced in the chapter. In this chapter, we only learned a few commands. We do not repeat the functions for the bioluminescent lattice plot and the penguin plot here, as these were used only for illustration. The functions discussed in this chapter are given in Table 1.1.

Table 1.1 R functions introduced in Chapter 1

Function	Purpose	Example
?	Access help files	`?boxplot`
#	Add comments	`#Add your comments here`
boxplot	Makes a boxplot	`boxplot (y)boxplot (y~factor (x))`
log	Natural logarithm	`log (2)`
log10	Logarithm with base 10	`log10 (2)`
library	Loads a package	`library (MASS)`
setwd	Sets the working directory	`setwd ("C:/AnyDirectory/")`
q	Closes R	`q()`
citation	Provides citation for R	`citation()`

Chapter 2
Getting Data into R

In the following chapter we address entering data into R and organising it as scalars (single values), vectors, matrices, data frames, or lists. We also demonstrate importing data from Excel, ascii files, databases, and other statistical programs.

2.1 First Steps in R

2.1.1 Typing in Small Datasets

We begin by working with an amount of data that is small enough to type into R. We use a dataset (unpublished data, Chris Elphick, University of Connecticut) containing seven body measurements taken from approximately 1100 saltmarsh sharp-tailed sparrows (*Ammodramus caudacutus*) (e.g., size of the head and wings, tarsus length, weight, etc.). For our purposes we use only four morphometric variables of eight birds (Table 2.1).

Table 2.1 Morphometric measurements of eight birds. The symbol NA stands for a missing value. The measured variables are the lengths of the wing (measured as the wing chord), leg (a standard measure of the tarsus), head (from the bill tip to the back of the skull), and weight.

Wingcrd	Tarsus	Head	Wt
59	22.3	31.2	9.5
55	19.7	30.4	13.8
53.5	20.8	30.6	14.8
55	20.3	30.3	15.2
52.5	20.8	30.3	15.5
57.5	21.5	30.8	15.6
53	20.6	32.5	15.6
55	21.5	NA	15.7

The simplest, albeit laborious, method of entering the data into R is to type it in as scalars (variables containing a single value). For the first five observations of wing length, we could type:

A.F. Zuur et al., *A Beginner's Guide to R*, Use R,
DOI 10.1007/978-0-387-93837-0_2, © Springer Science+Business Media, LLC 2009

```
> a <- 59
> b <- 55
> c <- 53.5
> d <- 55
> e <- 52.5
```

Alternatively, you can use the " = " symbol instead of " <−". If you type these commands into a text editor, then copy and paste them into R, nothing appears to happen. To see R's calculations, type "a" and click enter.

```
> a
[1] 59
```

Hence, "a" has the value of 59, as we intended. The problem with this approach is that we have a large amount of data and will quickly run out of characters. Furthermore, the variable names a, b, c, and so on are not very useful as aids for recalling what they represent. We could use variable names such as

```
> Wing1 <- 59
> Wing2 <- 55
> Wing3 <- 53.5
> Wing4 <- 55
> Wing5 <- 52.5
```

More names will be needed for the remaining data. Before we improve the naming process of the variables, we discuss what you can do with them. Once we have defined a variable and given it a value, we can do calculations with it; for example, the following lines contain valid commands.

```
> sqrt(Wing1)
> 2 * Wing1
> Wing1 + Wing2
> Wing1 + Wing2 + Wing3 + Wing4 + Wing5
> (Wing1 + Wing2 + Wing3 + Wing4 + Wing5) / 5
```

Although R performs the calculations, it does not store the results. It is perhaps better to define new variables:

```
> SQ.wing1 <- sqrt(Wing1)
> Mul.W1 <- 2 * Wing1
> Sum.12 <- Wing1 + Wing2
> SUM12345 <- Wing1 + Wing2 + Wing3 + Wing4 + Wing5
> Av <- (Wing1 + Wing2 + Wing3 + Wing4 + Wing5) / 5
```

These five lines are used to demonstrate that you can use any name. Note that the dot is a component of the name. We advise the use of variable names that aid in remembering what they represent. For example, SQ.wing1 reminds us that it is the square root of the wing length for bird 1. Sometimes, a bit of imagination is needed in choosing variable names. However, you should avoid names that contain characters like "£, $, %, ^ *, +, −, ( ), [ ], #, !, ?, <, >, and so on, as most of these characters are operators, for example, multiplication, power, and so on.

As we already explained above, if you have defined

```
> SQ.wing1 <- sqrt(Wing1)
```

to display the value of *SQ.wing1*, you need to type:

```
> SQ.wing1
[1] 7.681146
```

An alternative is to put round brackets around the command; R will now produce the resulting value:

```
> (SQ.wing1 <- sqrt(Wing1))
[1] 7.681146
```

2.1.2 Concatenating Data with the c Function

As mentioned above, with eight observations of four morphometric variables, we need 32 variable names. R allows the storage of multiple values within a variable. For this we need the c() function, where c stands for concatenate. It is used as follows.

```
> Wingcrd <- c(59, 55, 53.5, 55, 52.5, 57.5, 53, 55)
```

You may put spaces on either side of the commas to improve the readability of the code. Spaces can also be used on either side of the " + " and " <-" commands. In general, this improves readability of the code, and is recommended.

It is important to use the round brackets (and) in the c function and not the square [and] or the curly brackets { and }. These are used for other purposes.

Just as before, copying and pasting the above command into R only assigns the data to the variable Wingcrd. To see the data, type Wingcrd into R and press enter:

```
> Wingcrd
[1] 59.0 55.0 53.5 55.0 52.5 57.5 53.0 55.0
```

The c function has created a single vector of length 8. To view the first value of Wingcrd, type Wingcrd[1] and press enter:

```
> Wingcrd [1]
[1] 59
```

This gives the value 59. To view the first five values type:

```
> Wingcrd [1 : 5]
[1] 59.0 55.0 53.5 55.0 52.5
```

To view all except the second value, type:

```
> Wingcrd [-2]
[1] 59.0 53.5 55.0 52.5 57.5 53.0 55.0
```

Hence, the minus sign omits a value. R has many built-in functions, the most elementary of which are functions such as sum, mean, max, min, median, var, and sd, among othersn. They can be applied by typing

```
> sum(Wingcrd)
[1] 440.5
```

Obviously, we can also store the sum in a new variable

```
> S.win <- sum(Wingcrd)
> S.win
[1] 440.5
```

Again, the dot is part of the variable name. Now, enter the data for the other three variables from Table 2.1 into R. It is laborious, but typing the following code into an editor, then copying and pasting it into R does the job.

```
> Tarsus <- c(22.3, 19.7, 20.8, 20.3, 20.8, 21.5, 20.6,
           21.5)
> Head <- c(31.2, 30.4, 30.6, 30.3, 30.3, 30.8, 32.5,
           NA)
> Wt <- c(9.5, 13.8, 14.8, 15.2, 15.5, 15.6, 15.6,
           15.7)
```

Note that we are paying a price for the extra spaces; each command now extends into two lines. As long as you end the line with a backslash or a comma, R will consider it as one command.

It may be a good convention to capitalize variable names. This avoids confusion with existing function commands. For example, "head" is an existing function in R (see ? head). Most internal functions do not begin with a capital letter; hence we can be reasonably sure that Head is not an existing function. If you are not completely sure, try typing, for example,? Head. If a help file pops up, you know that you need to come up with another variable name.

Note that there is one bird for which the size of the head was not measured. It is indicated by NA. Depending on the function, the presence of an NA may, or may not, cause trouble. For example:

```
> sum(Head)
[1] NA
```

You will get the same result with the mean, min, max, and many other functions. To understand why we get NA for the sum of the head values, type ?sum. The following is relevant text from the sum help file.

```
...
sum(..., na.rm = FALSE)
...
If na.rm is FALSE, an NA value in any of the arguments
will cause a value of NA to be returned, otherwise NA
values are ignored.
...
```

Apparently, the default "na.rm = FALSE" option causes the R function sum to return an NA if there is a missing value in the vector (rm refers to remove). To avoid this, use "na.rm = TRUE"

```
> sum(Head, na.rm = TRUE)
[1] 216.1
```

Now, the sum of the seven values is returned. The same can be done for the mean, min, max, and median functions. On most computers, you can also use na.rm = T instead of na.rm = TRUE. However, because we have been confronted with classroom PCs running identical R versions on the same operating system, and a few computers give an error message with the na.rm = T option, we advise using na.rm = TRUE.

You should always read the help file for any function before use to ensure that you know how it deals with missing values. Some functions use na.rm, some use na.action, and yet others use a different syntax. It is nearly impossible to memorise how all functions treat missing values.

Summarising, we have entered data for four variables, and have applied simple functions such as mean, min, max, and so on. We now discuss methods of combining the data of these four variables: (1) the c, cbind, and rbind functions; (2) the matrix and vector functions; (3) data frames; and (4) lists.

Do Exercise 1 in Section 2.4 in the use of the c and sum functions.

2.1.3 Combining Variables with the c, cbind, and rbind Functions

We have four columns of data, each containing observations of eight birds. The variables are labelled Wingcrd, Tarsus, Head, and Wt. The c function was used to concatenate the eight values. In the same way as the eight values were concatenated, so can we concatenate the variables containing the values using:

```
> BirdData <- c(Wingcrd, Tarsus, Head, Wt)
```

Our use of the variable name BirdData instead of data, means that we are not overwriting an existing R function (see ? data). To see the result of this command, type BirdData and press enter:

```
> BirdData
 [1] 59.0 55.0 53.5 55.0 52.5 57.5 53.0 55.0 22.3
[10] 19.7 20.8 20.3 20.8 21.5 20.6 21.5 31.2 30.4
[19] 30.6 30.3 30.3 30.8 32.5   NA   9.5 13.8 14.8
[28] 15.2 15.5 15.6 15.6 15.7
```

BirdData is a single vector of length 32 (4 × 8). The numbers [1], [10], [19], and [28] are the index numbers of the first element on a new line. On your computer they may be different due to a different screen size. There is no need to pay any attention to these numbers yet.

R produces all 32 observations, including the missing value, as a single vector, because it does not distinguish values of the different variables (the first 8 observations are of the variable Wingcrd, the second 8 from Tarsus, etc.) . To counteract this we can make a vector of length 32, call it Id (for "identity"), and give it the following values.

```
> Id <- c(1, 1, 1, 1, 1, 1, 1, 1, 2, 2, 2, 2, 2, 2, 2,
    2, 3, 3, 3, 3, 3, 3, 3, 3, 4, 4, 4, 4, 4, 4, 4, 4)
> Id
 [1] 1 1 1 1 1 1 1 1 2 2 2 2 2 2 2 2 3 3 3 3 3 3 3
[24] 3 4 4 4 4 4 4 4 4
```

Because R can now put more digits on a line, as compared to in BirdData, only the indices [1] and [24] are produced. These indices are completely irrelevant for the moment. The variable Id can be used to indicate that all observations with a similar Id value belong to the same morphometric variable. However, creating such a vector is time consuming for larger datasets, and, fortunately, R has functions to simplify this process. What we need is a function that repeats the values 1 –4, each eight times:

```
> Id <- rep(c(1, 2, 3, 4), each = 8)
> Id
 [1] 1 1 1 1 1 1 1 1 2 2 2 2 2 2 2 2 3 3 3 3 3 3 3
[24] 3 4 4 4 4 4 4 4 4
```

This produces the same long vector of numbers as above. The rep designation stands for repeat. The command can be further simplified by using:

```
> Id <- rep(1 : 4, each = 8)
> Id
 [1] 1 1 1 1 1 1 1 1 2 2 2 2 2 2 2 2 3 3 3 3 3 3 3
[24] 3 4 4 4 4 4 4 4 4
```

Again, we get the same result. To see what the 1 : 4 command does, type into R:

```
> 1 : 4
```

It gives

```
[1] 1 2 3 4
```

So the : operator does not indicate division (as is the case with some other packages). You can also use the seq function for this purpose. For example, the command

```
> a <- seq(from = 1, to = 4, by = 1)
> a
```

creates the same sequence from 1 to 4,

```
[1] 1 2 3 4
```

So for the bird data, we could also use:

```
> a <- seq(from = 1, to = 4, by = 1)
> rep(a, each = 8)
 [1] 1 1 1 1 1 1 1 1 2 2 2 2 2 2 2 2 3 3 3 3 3 3 3
[24] 3 4 4 4 4 4 4 4 4
```

Each of the digits in "a" is repeated eight times by the rep function. At this stage you may well be of the opinion that in considering so many different options we are making things needlessly complicated. However, some functions in R need the data as presented in Table 2.1 (e.g, the multivariate analysis function for principal component analysis or multidimensional scaling), whereas the organisation of data into a single long vector, with an extra variable to identify the groups of observations (Id in this case), is needed for other functions such as the *t*-test, one-way anova, linear regression, and also for some graphing tools such as the xyplot in the lattice package (see Chapter 8). Therefore, fluency with the rep function can save a lot of time.

So far, we have only concatenated numbers. But suppose we want to create a vector "Id" of length 32 that contains the word "Wingcrd" 8 times, the word "Tarsus" 8 times, and so on. We can create a new variable called VarNames, containing the four morphometric variable designations. Once we have created it, we use the rep function to create the requested vector:

```
> VarNames <- c("Wingcrd", "Tarsus", "Head", "Wt")
> VarNames
[1] "Wingcrd" "Tarsus" "Head" "Wt"
```

Note that these are names, not the variables with the data values. Finally, we need:

```
> Id2 <- rep(VarNames, each = 8)
> Id2
 [1] "Wingcrd" "Wingcrd" "Wingcrd" "Wingcrd"
 [5] "Wingcrd" "Wingcrd" "Wingcrd" "Wingcrd"
 [9] "Tarsus"  "Tarsus"  "Tarsus"  "Tarsus"
[13] "Tarsus"  "Tarsus"  "Tarsus"  "Tarsus"
[17] "Head"    "Head"    "Head"    "Head"
[21] "Head"    "Head"    "Head"    "Head"
[25] "Wt"      "Wt"      "Wt"      "Wt"
[29] "Wt"      "Wt"      "Wt"      "Wt"
```

Id2 is a string of characters with the names in the requested order. The difference between Id and Id2 is just a matter of labelling. Note that you should not forget the "each =" notation. To see what happens if it is omitted, try typing:

```
> rep(VarNames, 8)
 [1] "Wingcrd" "Tarsus" "Head" "Wt"
 [5] "Wingcrd" "Tarsus" "Head" "Wt"
 [9] "Wingcrd" "Tarsus" "Head" "Wt"
[13] "Wingcrd" "Tarsus" "Head" "Wt"
[17] "Wingcrd" "Tarsus" "Head" "Wt"
[21] "Wingcrd" "Tarsus" "Head" "Wt"
[25] "Wingcrd" "Tarsus" "Head" "Wt"
[29] "Wingcrd" "Tarsus" "Head" "Wt"
```

It will produce a repetition of the entire vector VarNames with the four variable names listed eight times, not what we want in this case.

The c function is a way of combining data or variables. Another option is the cbind function. It combines the variables in such a way that the output contains the original variables in columns. For example, the output of the cbind function below is stored in Z. If we type Z and press enter, it shows the values in columns:

```
> Z <- cbind(Wingcrd, Tarsus, Head, Wt)
> Z
     Wingcrd   Tarsus   Head    Wt
[1,]    59.0     22.3   31.2   9.5
[2,]    55.0     19.7   30.4  13.8
[3,]    53.5     20.8   30.6  14.8
[4,]    55.0     20.3   30.3  15.2
[5,]    52.5     20.8   30.3  15.5
[6,]    57.5     21.5   30.8  15.6
[7,]    53.0     20.6   32.5  15.6
[8,]    55.0     21.5     NA  15.7
```

The data must be in this format if we are to apply, for example, principal component analysis. Suppose you want to access some elements of Z, for instance, the data in the first column. This is done with the command Z[, 1]:

```
> Z[, 1]
[1] 59.0 55.0 53.5 55.0 52.5 57.5 53.0 55.0
```

Alternatively, use

```
> Z[1 : 8, 1]
[1] 59.0 55.0 53.5 55.0 52.5 57.5 53.0 55.0
```

It gives the same result. The second row is given by Z[2,]:

```
> Z[2, ]
Wingcrd     Tarsus      Head       Wt
   55.0       19.7      30.4     13.8
```

Alternatively, you can use:

```
> Z[2, 1:4]
Wingcrd     Tarsus      Head       Wt
   55.0       19.7      30.4     13.8
```

The following commands are all valid.

```
> Z[1, 1]
> Z[, 2 : 3]
> X <- Z[4, 4]
> Y <- Z[, 4]
> W <- Z[, -3]
> D <- Z[, c(1, 3, 4)]
> E <- Z[, c(-1, -3)]
```

The first command accesses the value of the first bird for Wingcrd; the second command gives all the data for columns 2 and 3; X contains the weight for bird 4; and Y, all the Wt data. The minus sign is used to exclude columns or rows. Hence, W contains all variables except Head. We can also use the c function to access nonsequential rows or columns of Z. D contains the first, third, and fourth columns of Z, and E contains all but the first and third. You must ensure that the subscripts do not go outside the range of allowable values. For example, Z[8, 4] is valid, but Z[9, 5], Z[8, 6], or Z[10, 20] are not defined (we only have 8 birds and 4 variables). If you type one of these commands, R will give the error message:

```
Error: subscript out of bounds
```

If you would like to know the dimensions of Z, use:

```
> dim(Z)
[1] 8 4
```

The output is a vector with two elements: the number of rows and the number of columns of Z. At this point you may want to consult the help files of nrow and ncol for alternative options. In some situations, it may be useful to store the output of the dim function. In that case, use

```
> n <- dim(Z)
> n
 [1] 8 4
```

or, if you only need to store the number of rows in Z, use

```
> nrow <- dim(Z)[1]
> nrow
 [1] 8
```

Instead of nrow, the variable name zrow may be more appropriate. As you would expect, similar to the cbind function to arrange the variables in columns, the rbind function combines the data in rows. To use it, type:

```
> Z2 <- rbind(Wingcrd, Tarsus, Head, Wt)
> Z2
         [,1] [,2] [,3] [,4] [,5] [,6] [,7] [,8]
Wingcrd  59.0 55.0 53.5 55.0 52.5 57.5 53.0 55.0
Tarsus   22.3 19.7 20.8 20.3 20.8 21.5 20.6 21.5
Head     31.2 30.4 30.6 30.3 30.3 30.8 32.5   NA
Wt        9.5 13.8 14.8 15.2 15.5 15.6 15.6 15.7
```

This gives the same data as in the previous examples, with the morphometric variables in rows and the individual birds in columns.

Other interesting tools to change Z or Z2 are the edit and fix functions; see their help files.

Do Exercise 2 in Section 2.4 in the use of the c and cbind functions. This is an exercise using an epidemiological dataset.

2.1.4 Combining Data with the **vector** Function*

To avoid introducing too much information, we did not mention the vector function in the previous discussion, and upon first reading, you may skip this section. Instead of the c function, we could have used the vector function. Suppose we want to create a vector of length 8 containing data Wingcrd of all eight birds. In R, we can do this as follows.

```
> W <- vector(length = 8)
> W[1] <- 59
> W[2] <- 55
> W[3] <- 53.5
> W[4] <- 55
> W[5] <- 52.5
> W[6] <- 57.5
> W[7] <- 53
> W[8] <- 55
```

If you type W into R immediately after the first command, R shows a vector with values FALSE. This is normal. Typing W into R after all elements have been entered gives:

```
> W
[1] 59.0 55.0 53.5 55.0 52.5 57.5 53.0 55.0
```

Note that the result is identical to that of the c function. The advantage of the vector function is that we can define a priori how many elements a variable should have. This can be useful when doing specific tasks such as loops. However, for common applications, it is easier to use the c function to concatenate data.

Just as with the output of the c function, we can access particular elements of W using W[1], W[1 : 4], W[2 : 6], W[-2], W[c (1, 3, 5)], but W [9] produces an NA, as element 9 is not defined.

 Do Exercise 3 in Section 2.4 in the use of the vector function. This is an exercise using an epidemiological dataset.

2.1.5 Combining Data Using a Matrix*

Upon first reading, you may skip this section.

Instead of vectors showing the 4 variables Wingcrd, Tarsus, Head, and Wt, each of length 8, we can create a matrix of dimension 8 by 4 that contains the data. Such a matrix is created by the command:

```
> Dmat <- matrix(nrow = 8, ncol = 4)
> Dmat
        [,1]    [,2]    [,3]    [,4]
[1,]     NA      NA      NA      NA
[2,]     NA      NA      NA      NA
[3,]     NA      NA      NA      NA
[4,]     NA      NA      NA      NA
[5,]     NA      NA      NA      NA
[6,]     NA      NA      NA      NA
[7,]     NA      NA      NA      NA
[8,]     NA      NA      NA      NA
```

We first wanted to call this matrix D, but subsequently discovered that Tinn-R uses a blue font for D, meaning that it is an existing function. Entering ? D gives the information that it is a function to calculate derivates, hence we will not overwrite it. We instead use the designator "Dmat," where "mat" indicates matrix.

Note that Dmat is an 8 by 4 matrix containing only NAs. We need to fill in the values. This can be done by

```
> Dmat[, 1] <- c(59, 55, 53.5, 55, 52.5, 57.5, 53, 55)
> Dmat[, 2] <- c(22.3, 19.7, 20.8, 20.3, 20.8, 21.5,
                 20.6, 21.5)
> Dmat[, 3] <- c(31.2, 30.4, 30.6, 30.3, 30.3, 30.8,
                 32.5, NA)
> Dmat[, 4] <- c(9.5, 13.8, 14.8, 15.2, 15.5, 15.6,
                 15.6, 15.7)
```

The elements of Dmat, in this case, are entered by column, but we could have filled them in by row. Typing Dmat into R gives the same data matrix as we obtained with the cbind function, except that Dmat does not have column labels:

```
> Dmat
        [,1] [,2] [,3] [,4]
[1,] 59.0 22.3 31.2   9.5
[2,] 55.0 19.7 30.4 13.8
[3,] 53.5 20.8 30.6 14.8
[4,] 55.0 20.3 30.3 15.2
[5,] 52.5 20.8 30.3 15.5
[6,] 57.5 21.5 30.8 15.6
[7,] 53.0 20.6 32.5 15.6
[8,] 55.0 21.5   NA 15.7
```

We can use the existing `colnames` function to add column names to `Dmat`:

```
> colnames(Dmat) <- c("Wingcrd", "Tarsus", "Head","Wt")
> Dmat

     Wingcrd Tarsus Head   Wt
[1,]    59.0   22.3 31.2  9.5
[2,]    55.0   19.7 30.4 13.8
[3,]    53.5   20.8 30.6 14.8
[4,]    55.0   20.3 30.3 15.2
[5,]    52.5   20.8 30.3 15.5
[6,]    57.5   21.5 30.8 15.6
[7,]    53.0   20.6 32.5 15.6
[8,]    55.0   21.5   NA 15.7
```

Obviously, there is also a `rownames` function, the use of which is explained in the help file.

To summarise, we first defined a matrix of a specific size, then filled in its elements by column. You must define the matrix before you enter its elements. You can also fill in element by element, for example,

```
> Dmat[1, 1] <- 59.0
> Dmat[1, 2] <- 22.3
```

and so on, but this takes more time. If we have the data already categorized in variables, such as `Wingcrd`, `Tarsus`, `Head`, `Wt`, we would not normally create the matrix and fill in its elements. This command will do the job as well:

```
> Dmat2 <- as.matrix(cbind(Wingcrd, Tarsus, Head, Wt))
```

`Dmat2` and `Dmat` are identical. Once again learning more than one path to the same outcome is necessary because some functions require a matrix as input and will give an error message if a data frame (see next subsection) is used, and vice versa. Therefore, functions such as `as.matrix`, `is.matrix` (this function gives a TRUE if its argument is a matrix, and FALSE otherwise), `as.data.frame`, `is.date.frame` can come in handy.

Special operators for matrices A and B are `t(A)` for transpose, A %*% B for matrix multiplication, and `solve(A)` for inverse.

Do Exercise 4 in Section 2.4 dealing with matrices.

2.1.6 Combining Data with the **data.frame** Function

So far, we have used the c, cbind, rbind, vector, and matrix functions to combine data. Yet another option is the data frame. In a data frame we can combine variables of equal length, with each row in the data frame containing observations on the same sampling unit. Hence, it is similar to the matrix or cbind functions. Using the four bird morphometric variables from the previous section, a data frame is created as follows.

```
> Dfrm <- data.frame(WC = Wingcrd,
                      TS = Tarsus,
                      HD = Head,
                       W = Wt)
> Dfrm
    WC    TS    HD     W
1 59.0  22.3  31.2   9.5
2 55.0  19.7  30.4  13.8
3 53.5  20.8  30.6  14.8
4 55.0  20.3  30.3  15.2
5 52.5  20.8  30.3  15.5
6 57.5  21.5  30.8  15.6
7 53.0  20.6  32.5  15.6
8 55.0  21.5    NA  15.7
```

Basically, the data.frame function creates an object, called Dfrm in this case, and within Dfrm it stores values of the four morphometric variables. The advantage of a data frame is that you can make changes to the data without affecting the original data. For example, it is possible to combine the original (but renamed) weight and the square root transformed weights in the data frame Dfrm:

```
> Dfrm <- data.frame(WC = Wingcrd,
                      TS = Tarsus,
                      HD = Head,
                       W = Wt
                     Wsq = sqrt(Wt))
```

In the data frame, we can also combine numerical variables, character strings, and factors. Factors are nominal (categorical) variables and are discussed later in this chapter.

It is important to realise that the variable Wt that we created in the c function and the W in the data frame Dfrm are two different entities. To see this, let us remove the variable Wt (this is the one we typed in with the c function):

```
> rm(Wt)
```

If you now type in Wt, R gives an error message:

```
> Wt
Error: object "Wt" not found
```

But the variable W still exists in the data frame Dfrm:

```
> Dfrm$W
[1] 9.5 13.8 14.8 15.2 15.5 15.6 15.6 15.7
```

It may seem that the data frame is unnecessary, because we have the cbind-dand matrix functions, However, neither of these can be used to combine different types of data. Our use of the data frame is often as follows. First we enter the data into R, mainly using methods discussed in Section 2.2. We then make changes to the data (e.g., remove extreme observations, apply transformations, add categorical variables, etc.) and store the modified data in a data frame which we use in the follow-up analyses.

2.1.7 Combining Data Using the *list* Function*

You may also skip this section at first reading. So far, the tools we have used to combine data produce a table with each row being a sampling unit (a bird in this case). Suppose you want a black box into which you can put as many variables as you want; some may be related, some may have similar dimensions, some may be vectors, others matrices, and yet others may contain character strings of variable names. This is what the listfunction can do. The main difference from our previously used methods is that the resulting rows will not necessarily represent a single sampling unit. A simple example is given below. The variables x1, x2, x3, and x4 contain data: x1 is a vector of length 3, x2 contains 4 characters, x3 is a variable of dimension 1, and x4 is a matrix of dimension 2-by-2. All these variables are entered into the list function:

```
> x1 <- c(1, 2, 3)
> x2 <- c("a", "b", "c", "d")
> x3 <- 3
> x4 <- matrix(nrow = 2, ncol = 2)
> x4[, 1] <- c(1, 2)
> x4[, 2] <- c( 3, 4)
> Y <- list(x1 = x1, x2 = x2, x3 = x3, x4 = x4)
```

If you now type Y into R, you get the following output.

```
> Y

$x1
[1] 1 2 3

$x2
[1] "a" "b" "c" "d"

$x3
[1] 3

$x4
      [,1]  [,2]
[1,]    1     3
[2,]    2     4
```

All information contained in Y is accessible by typing, for example, Y$x1, Y$x2, and so on. Again, you may wonder why we need to go to all this trouble. The reason is that nearly all functions (e.g., linear regression, generalised linear modelling, *t*-test, etc.) in R produce output that is stored in a list. For example, the following code applies a linear regression model in which wing length is modelled as a function of weight.

```
> M <- lm(WC ~ Wt, data = Dfrm)
```

We do not wish to go into detail of the lm function, or how to do linear regression in R (see its helpfile obtained by typing ? lm). All what we want to emphasise is that R stores all results of the linear regression function in the object M. If you type

```
> names (M)
```

you receive this fancy output:

```
 [1] "coefficients"   "residuals"       "effects"
 [4] "rank"           "fitted.values"   "assign"
 [7] "qr"             "df.residual"     "xlevels"
[10] "call"           "terms"           "model"
```

You can access the coefficients or residuals by using M$coefficients, M$residuals, and so on. Hence, M is a list containing objects of different dimensions, just as our earlier example with Y. The good news is that R contains various functions to extract required information (e.g., estimated values, *p*-values, etc.) and presents it in nice tables. See the lm help file.

For the bird morphometric data, it does not make sense to store the data in a list, as the rows in Table 2.1 contain data from the *same* bird. However, if the task is to create a list that contains all data in a long vector, an extra vector that identifies the groups of variables (ID in this case), a matrix that contains the data in a 8 by 4 format, and, finally, a vector that contains the 4 morphometric names, we can use:

```
> AllData <- list(BirdData = BirdData, Id = Id2, Z = Z,
                  VarNames = VarNames)
```

to produce:

```
> AllData
$BirdData
 [1] 59.0 55.0 53.5 55.0 52.5 57.5 53.0 55.0 22.3
[10] 19.7 20.8 20.3 20.8 21.5 20.6 21.5 31.2 30.4
[19] 30.6 30.3 30.3 30.8 32.5   NA  9.5 13.8 14.8
[28] 15.2 15.5 15.6 15.6 15.7

$Id
 [1] "Wingcrd" "Wingcrd" "Wingcrd" "Wingcrd"
 [5] "Wingcrd" "Wingcrd" "Wingcrd" "Wingcrd"
 [9] "Tarsus"  "Tarsus"  "Tarsus"  "Tarsus"
[13] "Tarsus"  "Tarsus"  "Tarsus"  "Tarsus"
[17] "Head"    "Head"    "Head"    "Head"
[21] "Head"    "Head"    "Head"    "Head"
[25] "Wt"      "Wt"      "Wt"      "Wt"
[29] "Wt"      "Wt"      "Wt"      "Wt"

$Z
     Wingcrd Tarsus Head   Wt
[1,]    59.0   22.3 31.2  9.5
[2,]    55.0   19.7 30.4 13.8
[3,]    53.5   20.8 30.6 14.8
[4,]    55.0   20.3 30.3 15.2
[5,]    52.5   20.8 30.3 15.5
[6,]    57.5   21.5 30.8 15.6
[7,]    53.0   20.6 32.5 15.6
[8,]    55.0   21.5   NA 15.7

$VarNames
[1]"Wingcrd""Tarsus" "Head" "Wt"
```

Obviously, storing the data in this format is unnecessary, as we only need one format. An advantage, perhaps, with multiple formats, is that we are prepared for most functions. However, our own programming style is such that we only change the format if, and when, needed.

Typing `AllData` in R produces the data in most formats that we have seen in this section. It is nice to know that we can do it.

You cannot use the " < -" symbols in the `list` function, only the " = " sign is accepted. Figure 2.1 shows an overview of the methods of storing data discussed so far.

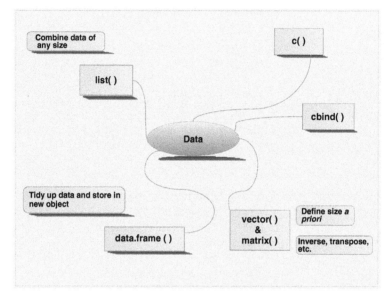

Fig. 2.1 Overview of various methods of storing data. The data stored by `cbind`, `matrix`, or `data.frame` assume that data in each row correspond to the same observation (sample, case)

Do Exercise 5 in Section 2.4. This is an exercise that deals again with an epidemiological dataset and the use of the `data.frame` and `list` commands.

2.2 Importing Data

With large datasets, typing them in, as we did in the previous section, is not practical. The most difficult aspect of learning to use a new package is importing your data. Once you have mastered this step, you can experiment with other commands. The following sections describe various options for importing data. We make a distinction between small and large datasets and whether they are stored in Excel, ascii text files, a database program, or in another statistical package.

2.2.1 Importing Excel Data

There are two main options for importing data from Excel (or indeed any spreadsheet or database program) into R. The easy, and recommended, option is (1) to prepare the data in Excel, (2) export it to a tab-delimited ascii file, (3) close Excel, and (4) use the read.table function in R to import the data. Each of these steps is discussed in more detail in the following sections. The second option is a special R package, RODBC, which can access selected rows and columns from an Excel spreadsheet. However, this option is not for the fainthearted. Note that Excel is not particularly suitable for working with large datasets, due to its limitation in the number of columns.

2.2.1.1 Prepare the Data in Excel

In order to keep things simple, we recommend that you arrange the data in a sample-by-variable format. By this, we mean with the columns containing variables, and the rows containing samples, observations, cases, subjects, or whatever you call your sampling units. Enter NA (in capitals) into cells representing missing values. It is good practice to use the first column in Excel for identifying the sampling unit, and the first row for the names of the variables. As mentioned earlier, using names containing symbols such as £, $, %, ^, &, *, (,), −, #, ?, , ,.. , <, >, /, |, \, ,[,] ,{, and } will result in an error message in R. You should also avoid names (or fields or values) that contain spaces. Short names are advisable in order to avoid graphs containing many long names, making the figure unreadable.

Figure 2.2 shows an Excel spreadsheet containing a set of data on the Gonadosomatic index (GSI, i.e., the weight of the gonads relative to total body weight) of squid (Graham Pierce, University of Aberdeen, UK, unpublished data). Measurements were taken from squid caught at various locations in Scottish waters in different months and years.

2.2.1.2 Export Data to a Tab-Delimited ascii File

In Excel, go to **File- > Save As- > Save as Type**, and select **Text (Tab delimited)**. If you have a computer running in a non-English language, it may be a challenge to determine how "Tab delimited" is translated. We exported the squid data in Fig. 2.2 to a tab-delimited ascii file named squid.txt, in the directory C:\RBook. Both the Excel file and the tab-delimited ascii file can be downloaded from the book's website. If you download them to a different directory, then you will need to adjust the "C:\RBook" path.

At this point it is advisable to close Excel so that it cannot block other programs from accessing your newly created text file.

Warning: Excel has the tendency to add extra columns full of NAs to the ascii file if you have, at some stage, typed comments into the spreadsheet. In R, these columns may appear as NAs. To avoid this, delete such columns in Excel before starting the export process.

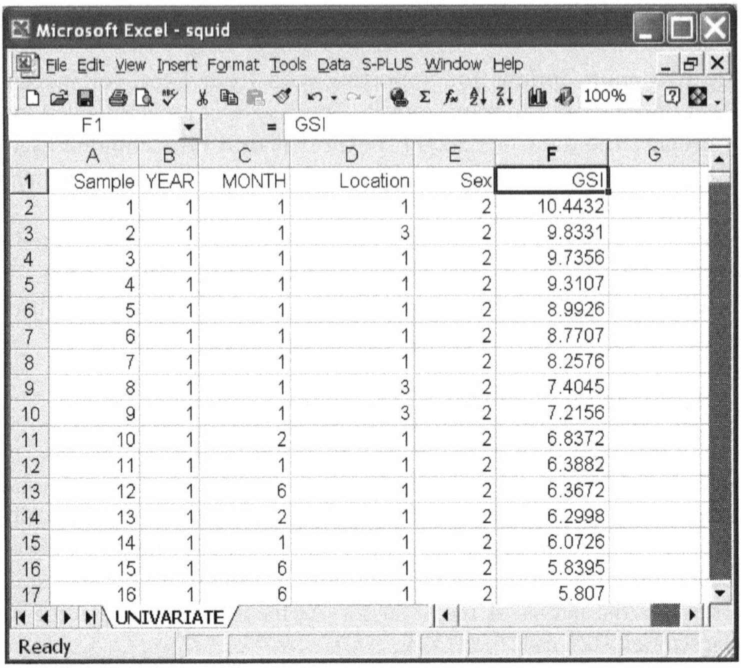

	A	B	C	D	E	F	G
1	Sample	YEAR	MONTH	Location	Sex	GSI	
2	1	1	1	1	2	10.4432	
3	2	1	1	3	2	9.8331	
4	3	1	1	1	2	9.7356	
5	4	1	1	1	2	9.3107	
6	5	1	1	1	2	8.9926	
7	6	1	1	1	2	8.7707	
8	7	1	1	1	2	8.2576	
9	8	1	1	3	2	7.4045	
10	9	1	1	3	2	7.2156	
11	10	1	2	1	2	6.8372	
12	11	1	1	1	2	6.3882	
13	12	1	6	1	2	6.3672	
14	13	1	2	1	2	6.2998	
15	14	1	1	1	2	6.0726	
16	15	1	6	1	2	5.8395	
17	16	1	6	1	2	5.807	

Fig. 2.2 Example of the organisation of a dataset in Excel prior to importing it into R. The rows contain the cases (each row represents an individual squid) and the columns the variables. The first column and the first row contain labels, there are no labels with spaces, and there are no empty cells in the columns that contain data

2.2.1.3 Using the `read.table` Function

With a tab-delimited ascii file that contains no blank cells or names with spaces, we can now import the data into R. The function that we use is `read.table`, and its basic use is as follows.

```
> Squid <- read.table(file = "C:\\RBook\\squid.txt",
                      header = TRUE)
```

This command reads the data from the file *squid.txt* and stores the data in `Squid` as a data frame. We highly recommend using short, but clear, variable labels. For example, we would not advise using the name `SquidNorthSea-MalesFemales`, as you will need to write out this word frequently. A spelling mistake and R will complain. The `header = TRUE` option in the `read.table` function tells R that the first row contains labels. If you have a file without headers, change it to `header = FALSE`There is another method of specifying the location of the text file:

```
> Squid <- read.table(file = "C:/RBook/squid.txt",
                      header = TRUE)
```

Note the difference in the slashes. If you have error messages at this stage, make sure that the file name and directory path were correctly specified. We strongly advise keeping the directory names simple. We have seen too many people struggling for half an hour to get the `read.table` function to run when they have a typing error in the 150–200-character long directory structure. In our case, the directory structure is short, C:/RBook. In most cases, the directory path will be longer. It is highly likely that you will make a mistake if you type in the directory path from memory. Instead, you can right-click the file Squid.txt (in Windows Explorer), and click Properties (Fig. 2.3). From here, you can copy and paste the full directory structure (and the file name) into your R text editor. Don't forget to add the extra slash \.

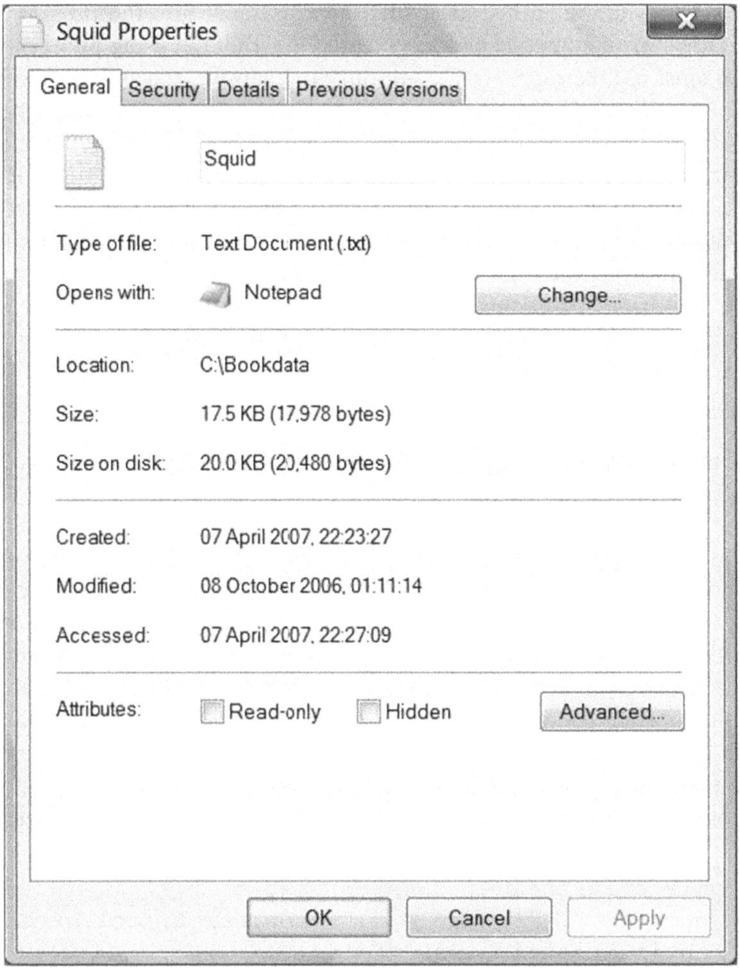

Fig. 2.3 Properties of the file *squid.txt*. The file name is Squid.txt, and the location is C:\Bookdata. You can highlight the location, copy it, paste it into the `read.table` function in your text R editor, and add the extra \ on Windows operating systems

If you use names that include "My Files," be sure to include the space and the capitals. Another common reason for an error message is the character used for decimal points. By default, R assumes that the data in the ascii text file have point separation, and the read.table function is actually using:

```
> Squid <- read.table(file = "C:/RBook/squid.txt",
              header = TRUE, dec = ".")
```

If you are using comma separation, change the last option to `dec = ","`, and rerun the command.

Warning: If your computer uses comma separation, and you export the data from Excel to a tab-delimited ascii file, then you must use the `dec = ","` option. However, if someone else has created the ascii file using point separation, you must use the `dec = "."` option. For example, the ascii files for this book are on the book website and were created with point separation. Hence all datasets in this book have to be imported with the `dec = "."` option, even if your computer uses comma separation. If you use the wrong setting, R will import all numerical data as categorical variables. In the next chapter, we discuss the str function, and recommend that you always execute it immediately after importing the data to verify the format.

If the data have spaces in the variable names, and if you use the read.-table function as specified above, you will get the following message. (We temporarily renamed GSI to G S I in Excel in order to create the error message.)

```
Error in scan(file,what,nmax,sep,dec,quote,skip,nlines,
na.strings,: line 1 did not have 8 elements
```

R is now complaining about the number of elements per line. The easy option is to remove any spaces from names or data fields in Excel and repeat the steps described above. The same error is obtained if the data contain an empty cell or fields with spaces. Instead of modifying the original Excel file, it is possible to tell the read.table function that there will be fields with spaces. There are many other options for adapting the read.table function to your data. The best way to see them all is to go to the read.table help file. The first part of the help file is shown below. You are not expected to know the meaning of all the options, but it is handy to know that it is possible to change certain settings.

```
read.table(file, header = FALSE, sep = "",
       quote = "\"'", dec = ".", row.names, col.names,
       as.is = !stringsAsFactors,
       na.strings = "NA", colClasses = NA, nrows=-1,
       skip = 0, check.names = TRUE,
       fill = !blank.lines.skip,
```

```
             strip.white = FALSE, blank.lines.skip = TRUE,
             comment.char = "#", allowEscapes = FALSE,
             flush = FALSE,
             stringsAsFactors = default.stringsAsFactors())
```

This is a function with many options. For example if you have white space in the fields, use the option `strip.white = TRUE`. An explanation of the other options can be found under the Arguments section in the help file. The help file also gives information on reading data in csv format. It is helpful to know that the `read.table` can contain an URL link to a text file on an Internet webpage.

If you need to read multiple files from the same directory, it is more efficient (in terms of coding) to set the working directory with the `setwd` function. You can then omit the directory path in front of the text file in the `read.table` function. This works as follows.

```
> setwd("C:\\RBook\\")
> Squid <- read.table(file = "squid.txt",
             header = TRUE)
```

In this book, we import all datasets by designating the working directory with the `setwd` function, followed by the `read.table` function. Our motivation for this is that not all users of this book may have permission to save files on the C drive (and some computers may not have a C drive!). Hence, they only need to change the directory in the `setwd` function.

In addition to the `read.table` function, you can also import data with the `scan` function. The difference is that the `read.table` stores the data in a data frame, whereas the `scan` function stores the data as a matrix. The `scan` function will work faster (which is handy for large datasets, where large refers to millions of data points), but all the data must be numerical. For small datasets, you will hardly know the difference in terms of computing time. For further details on the `scan` function, see its help file obtained with `? scan`.

 Do Exercises 6 and 7 in Section 2.4 in the use of the `read.table` and `scan` functions. These exercises use epidemiological and deep sea research data.

*2.2.2 Accessing Data from Other Statistical Packages***

In addition to accessing data from an ascii file, R can import data from other statistical packages, for example, Minitab, S-PLUS, SAS, SPSS, Stata, and Systat, among others. However, we stress that it is best to work with the original data directly, rather than importing data possibly modified by another statistical software package. You need to type:

```
> library(foreign)
```

in order to access these options. The help file for reading Minitab files is obtained by typing:

```
> ?read.mtp
```

and provides a good starting point. There is even a write.foreign function with the syntax:

```
write.foreign(df, datafile, codefile,
     package = c("SPSS", "Stata", "SAS"), ...)
```

Hence, you can export information created in R to some of the statistical packages. The options in the function write.foreign are discussed in its help file.

2.2.3 Accessing a Database***

This is a rather more technical section, and is only relevant if you want to import data from a database. Accessing or importing data from a database is relatively easy. There is a special package available in R that provides the tools you need to quickly gain access to any type of database. Enter:

```
> library(RODBC)
```

to make the necessary objects available. The package implements Open DataBase Connectivity (ODBC) with compliant databases when drivers exist on the host system. Hence, it is essential that the necessary drivers were set up when installing the database package. In Windows, you can check this through the Administrative Tools menu or look in the Help and Support pages under ODBC. Assuming you have the correct drivers installed, start by setting up a connection to a Microsoft Access database using the odbcConnectAccess: command. Let us call that connection channel1; so type in:

```
> setwd("C:/RBook")
> channel1 <- odbcConnectAccess(file =
     "MyDb.mdb", uid = "", pwd = "")
```

As you can see, the database, called MyDB.mdb, does not require a user identification (uid) or password (pwd) which, hence, can be omitted. You could have defined a database on your computer through the DSN naming protocol as shown in Fig. 2.4.

Now we can connect to the database directly using the name of the database:

```
> Channel1 <- odbcConnect("MyDb.mdb")
```

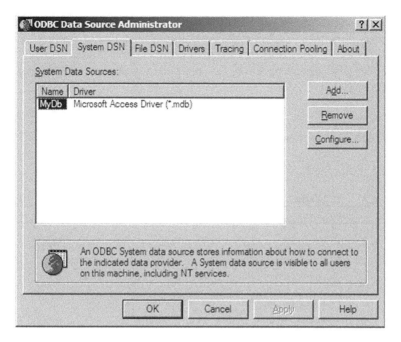

Fig. 2.4 Windows Data Source Administrator with the database MyDb added to the system data source names

Once we have set up the connection it is easy to access the data in a table:

```
> MyData <- sqlFetch(channel1, "MyTable")
```

We use `sqlFetch` to fetch the data and store it in `MyData`. This is not all you can do with an ODBC connection. It is possible to select only certain rows of the table in the database, and, once you have mastered the necessary database language, to do all kinds of fancy queries and table manipulations from within R. This language, called Structured Query Language, or SQL, is not difficult to learn. The command used in RODBC to send an SQL query to the database is `sqlQuery(channel, query)` in which `query` is simply an SQL query between quotes. However, even without learning SQL, there are some commands available that will make working with databases easier. You can use `sqlTables` to retrieve table information in your database with the command `SqlTables(channel)` or `sqlColumns(channel, "MyTable")` to retrieve information in the columns in a database table called MyTable. Other commands are `sqlSave`, to write or update a table in an ODBC database; `sqlDrop`, to remove a table; and `sqlClear`, to delete the content.

Windows users can use odbcConnectExcel to connect directly to Excel spreadsheet files and can select rows and columns from any of the sheets in the file. The sheets are represented as different tables.

There are also special packages for interfacing with Oracle (ROracle) and MySQL (RMySQL).

2.3 Which R Functions Did We Learn?

Table 2.2 shows the R functions introduced in this chapter.

Table 2.2 R functions introduced in this chapter

Function	Purpose	Example
sum	Calculated the sum	sum (x, na.rm = TRUE)
median	Calculated the median	median (x, na.rm = TRUE)
max	Calculated the maximum	max (x, na.rm = TRUE)
min	Calculated the minimum	min (x, na.rm = TRUE)
c()	Concatenate data	c (1, 2, 3)
cbind	Combine variables in columns	cbind (x, y, z)
rbind	Combine variables in rows	rbind (x, y, z)
vector	Combine data in a vector	vector (length = 10)
matrix	Combine data in a matrix	matrix (nrow = 5, ncol = 10)
data.frame	Combine data in a data frame	data.frame (x = x, y = y, z = z)
list	Combine data in a list	list (x = x, y = y, z = z)
rep	Repeat values or variables	rep (c (1, 2, 3), each = 10)
seq	Create a sequence of numbers	seq (1, 10)
dim	Dimension of a matrix or cbind output	dim (MyData)
colnames	Column names of a matrix or cbind output	colnames (MyData)
rownames	Row names of a matrix or cbind output	rownames (MyData)
setwd	Sets the working directory	setwd ("C:/Rbook/")
read.table	Read data from an ascii file	read.table (file = "test.txt", header = TRUE)
scan	Read data from an ascii file	scan (file ="test.txt")

2.4 Exercises

Exercise 1. The use of the c and sum functions.
This exercise uses epidemiological data. Vicente et al. (2006) analysed data from observations of wild boar and red deer reared on a number of estates in Spain. The dataset contains information on tuberculosis (Tb) in both species, and on the parasite *Elaphostrongylus cervi*, which only infects red deer.

In Zuur et al. (2009), Tb was modelled as a function of the continuous explanatory variable, length of the animal, denoted by LengthCT (CT is an abbreviation of *cabeza-tronco*, which is Spanish for head-body). Tb and Ecervi are shown as a vector of zeros and ones representing absence or presence of Tb and *E. cervi* larvae. Below, the first seven rows of the spreadsheet containing the deer data are given.

Farm	Month	Year	Sex	LengthClass	LengthCT	Ecervi	Tb
MO	11	00	1	1	75	0	0
MO	07	00	2	1	85	0	0
MO	07	01	2	1	91.6	0	1
MO	NA	NA	2	1	95	NA	NA
LN	09	03	1	1	NA	0	0
SE	09	03	2	1	105.5	0	0
QM	11	02	2	1	106	0	0

Using the c function, create a variable that contains the length values of the seven animals. Also create a variable that contains the Tb values. Include the NAs. What is the average length of the seven animals?

Exercise 2. The use of the cbind function using epidemiological data.

We continue with the deer from Exercise 1. First create variables Farm and Month that contain the relevant information. Note that Farm is a string of characters. Use the cbind command to combine month, length, and Tb data, and store the results in the variable, Boar. Make sure that you can extract rows, columns, and elements of Boar. Use the dim, nrow, and ncol functions to determine the number of animals and variables in Boar.

Exercise 3. The use of the vector function using epidemiological data.

We continue with the deer from Exercise 1. Instead of the cfunction that you used in Exercise 2 to combine the Tb data, can you do the same with the vectorfunction? Give the vector a different name, for example, Tb2.

Exercise 4. Working with a matrix.

Create the following matrix in R and determine its transpose, its inverse, and multiple D with its inverse (the outcome should be the identity matrix).

$$D = \begin{pmatrix} 1 & 2 & 3 \\ 4 & 2 & 1 \\ 2 & 3 & 0 \end{pmatrix}$$

Exercise 5. The use of the data.frame and list functions using epidemiological data.

We continue with the deer from Exercises 1 to 3. Make a data frame that contains all the data presented in the table in Exercise 1. Suppose that you decide to square root transform the length data. Add the square root transformed data to the data frame. Do the same exercise with a `list` instead of a `data.frame`. What are the differences?

Exercise 6. The use of the `read.table` and `scan` functions using deep sea research data.

The file ISIT.xls contains the bioluminescent data that were used to make Fig. 1.6. See the paragraph above this graph for a description. Prepare the spreadsheet (there are 4–5 problems you will need to solve) and export the data to an ascii file. Import the data into R using first the `read.table` function and then the `scan` function. Use two different names under which to store the data. What is the difference between them? Use the `is.matrix` and `is.data.frame` functions to answer this question.

Exercise 7. The use of the `read.table` or `scan` function using epidemiological data.

The file *Deer.xls* contains the deer data discussed in Exercise 1, but includes all animals. Export the data in the Excel file to an ascii file, and import it into R.

Chapter 3
Accessing Variables and Managing Subsets of Data

In the previous chapter we demonstrated importing data from a spreadsheet or database into R. We also showed how to type in small datasets and store them in a data frame. We now discuss accessing subsets of the data.

3.1 Accessing Variables from a Data Frame

Assuming that no errors were encountered when importing the squid data in the previous section, we can now move on to working with the data.

During statistical analysis it can be important to remove portions of the data, select certain subsets, or sort them. Most of these operations can be done in Excel or other spreadsheet (or database) program prior to importing into R, but, for various reasons, it is best not to do this. You may end up needing to reimport your data each time you make a subselection. Also some data files may be too large to import from a spreadsheet. Hence, a certain degree of knowledge of manipulating data files in R is useful. However, the reader is warned that this may be the most difficult aspect of R, but once mastered it is rewarding, as it means that all the tedious data manipulation in Excel (or any other spreadsheet) can be done in R.

We use the squid data imported in the previous chapter. If you have not already done this, use the following commands to import the data and store it in the data frame Squid.

```
> setwd("C:/RBook/")
> Squid <- read.table(file = "squid.txt",
                header = TRUE)
```

The read.table function produces a data frame, and, because most functions in R work with data frames, we prefer it over the scan function. We advise using the names command immediately after the read.table command to see the variables we are dealing with:

```
> names(Squid)
[1] "Sample" "Year" "Month" "Location" "Sex" "GSI"
```

We often notice that our course participants continue typing code directly into the R console. As mentioned in Chapter 1, we strongly recommend typing commands into a good text editor, such as Tinn-R on Windows operating systems. (See Chapter 1 for sources of editors using non-Windows operating systems.) To emphasise this, Fig. 3.1 shows a snapshot of our R code so far. Note that we copied and pasted the results of the `names` command back into our Tinn-R file. This allows us to see quickly which variables we will work with, and reduces the chance of typing errors.

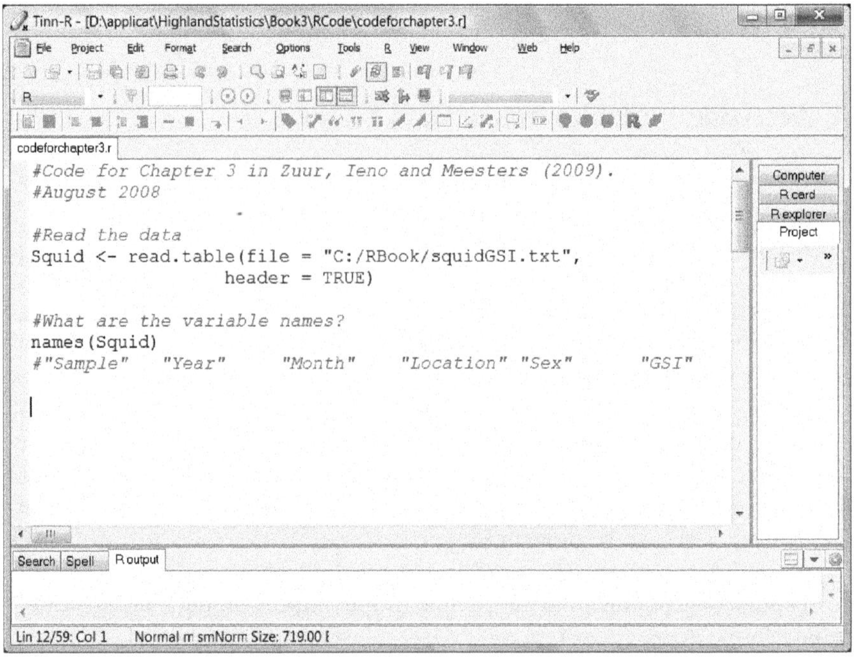

Fig. 3.1 Snapshot of our Tinn-R file. Note that the "#" symbol is put before comments, and that the code is well documented, including the date when it was written. Copying and pasting all variable names into the text file allows us to quickly check the spelling of the variable names. It is important that you structure your file as transparently as possible and add comments. Also ensure that you have a backup of this file and the data file

3.1.1 The str Function

The str (structure) command informs us of the status of each variable in a data frame:

```
> str(Squid)
'data.frame': 2644 obs. of 6 variables:
 $ Sample   : int 1 2 3 4 5 6 7 8 9 10 ...
 $ Year     : int 1 1 1 1 1 1 1 1 1 1 ...
 $ Month    : int 1 1 1 1 1 1 1 1 1 2 ...
 $ Location : int 1 3 1 1 1 1 1 3 3 1 ...
 $ Sex      : int 2 2 2 2 2 2 2 2 2 2 ...
 $ GSI      : num 10.44 9.83 9.74 9.31 8.99 ...
```

This cryptic output tells us that the variables Sample, Year, Month, Location, and Sex are integers, and GSI is numeric. Suppose that you made a mistake with the point separation:

```
> setwd("C:/RBook/")
> Squid2 <- read.table(file = "squidGSI.txt",
                       dec = ",", header = TRUE)
```

We (wrongly) told R that the decimal separation in the ascii file is a comma. The Squid2 data frame still contains the same data, but using the str command allows us to detect a major problem:

```
> str(Squid2)
'data.frame': 2644 obs. of 6 variables:
 $ Sample   : int 1 2 3 4 5 6 7 8 9 10 ...
 $ Year     : int 1 1 1 1 1 1 1 1 1 1 ...
 $ Month    : int 1 1 1 1 1 1 1 1 1 2 ...
 $ Location : int 1 3 1 1 1 1 1 3 3 1 ...
 $ Sex      : int 2 2 2 2 2 2 2 2 2 2 ...
 $ GSI      : Factor w/ 2472 levels "0.0064","0.007" ...
```

The GSI variable is now considered to be a factor. This means that if, in continuing, we use functions such as the mean or a boxplot, R will produce cryptic error messages, as GSI is not numerical. We have seen a lot of confusion due to this type of mistake.

Therefore we strongly recommend that you always combine the read.table function with the names and str functions.

The variable of interest is GSI, and, in any follow-up statistical analysis, we may want to model it as a function of year, month, location, and sex. Before doing any statistical analysis, one should always visualise the data (i.e., make plots). Useful tools are boxplots, Cleveland dotplots, scatterplots, pair plots, and the like (see Zuur et al., 2007; 2009). However, R does not recognize the variable GSI (or any of the other variables). To demonstrate this, type

```
> GSI
Error: object "GSI" not found
```

The problem is that the variable GSI is stored in the data frame Squid. There are several ways to access it, both good ways and bad ways, and we discuss them next.

3.1.2 The Data Argument in a Function

The most efficient method of accessing variables in a data frame is as follows. Identify a function in R, for instance, lm for linear regression; specify the model in terms of the variables GSI, Month, Year, and Location; and tell the function lm that the data can be found in the data frame Squid. Although we are not further discussing linear regression in this book, the code would look as follows.

```
> M1 <- lm(GSI ~ factor(Location) + factor(Year),
            data = Squid)
```

We ignore the first part, which specifies the actual linear regression model. It is the last part of the statement (data =) which tells R that the variables are in the data frame Squid. This is a neat approach, as there is no need to define variables outside the data frame; everything is nicely stored in a data frame Squid. The major problem with this approach is that not all functions support the data option. For example,

```
> mean(GSI, data = Squid)
```

gives an error message:

```
Error in mean (GSI, data = Squid) : object "GSI" not
found
```

because the function mean does not have a data argument. And sometimes a help file tells you that there is a data argument, and it may work in some cases, but not in other cases. For example, the code below gives a boxplot (not shown here).

```
> boxplot(GSI ~ factor(Location), data = Squid)
```

But this command gives an error message:

```
> boxplot(GSI, data = Squid)
Error in boxplot(GSI, data = Squid) : object "GSI" not
found
```

To summarise, if a function has a `data` argument, use it; it is the neatest programming approach.

3.1.3 The $ Sign

So, what can you do if a function does not have a data argument? There are two ways to access a variable. The first option is the $ sign:

```
> Squid$GSI
 [1] 10.4432   9.8331   9.7356   9.3107   8.9926
 [6]  8.7707   8.2576   7.4045   7.2156   6.8372
[11]  6.3882   6.3672   6.2998   6.0726   5.8395
```

< Cut to reduce space >

We only copied and pasted the first few lines of the output as the dataset contains 2644 observations. The other variables can be accessed in the same way. Type the name of the data frame followed by a $ and the name of the variable. In principle, you can put spaces between the $ sign and the variable names:

```
> Squid$GSI
 [1] 10.4432 9.8331 9.7356 9.3107 8.9926
 [6] 8.7707 8.2576 7.4045 7.2156 6.8372
[11] 6.3882 6.3672 6.2998 6.0726 5.8395
```

< Cut to reduce space >

We do not recommend this (it looks odd).
The second approach is to select the sixth column if you want to access the GSI data:

```
> Squid[, 6]
 [1] 10.4432 9.8331 9.7356 9.3107 8.9926
 [6] 8.7707 8.2576 7.4045 7.2156 6.8372
[11] 6.3882 6.3672 6.2998 6.0726 5.8395
```

< Cut to reduce space >

It gives exactly the same result. Using either `Squid$GSI` or `Squid[, 6]`, you can now calculate the mean:

```
> mean(Squid$GSI)
[1] 2.187034
```

Our preference is the coding with $GSI. A week after you typed Squid[, 6],
you will have forgotten that the GSI data are in the sixth column, and the
notation $GSI is clearer.

To add to the confusion, you can also use Squid[, "GSI"]. Note that in some
functions, the use of the Squid$ approach gives an error message, for example,
the gls function from the nlme package.

3.1.4 The attach Function

Let us now discuss a bad way of accessing variables. We have used "$" to access
variables from the data frame Squid. It can be tedious typing Squid$ each
time we want to use certain variables from the GSI dataset. It is possible to
avoid this by using the attach command. This command makes all variables
inside the data frame Squid available. To be more precise, the attach
command adds Squid to the search path of R. As a result, you can now type
GSI or Location without using Squid$.

```
> attach(Squid)
> GSI
 [1] 10.4432   9.8331   9.7356   9.3107   8.9926
 [6]  8.7707   8.2576   7.4045   7.2156   6.8372
[11]  6.3882   6.3672   6.2998   6.0726   5.8395
```

< Cut to reduce space >

The same holds for the other variables. As a result, you can now use each
function without a data argument:

```
> boxplot(GSI) #Graph not shown here
> mean(GSI)
[1] 2.187034
```

The attach command sounds too good to be true. It can be a useful
command, if used with great care. Problems occur if you attach a dataset that
has variable names that also exist outside the data frame. Or if you attach two
data frames, and variables with the same names appear in both. Another problem
may occur if you attach a dataset that has variable names that match some of R's
own function names or names of variables in example datasets (e.g., the variable
name "time" and the function "time"). In all these cases, you may discover that R
will not use the variable in your calculations as you expected. This is a major
problem in classroom teaching when students do different exercises and each time
load a new dataset with similar names such as "Location," "Month," "Sex," and
so on. In such situations it is better to use the detach command, or simply close
and restart R each time you work with a new dataset. If you use only one dataset
for a research project and are careful with variable names, the attach command
is extremely useful. But use it with care.

To summarise the use of the `attach` command,

1. To avoid duplicate variables, do not type the `attach` (`Squid`) command twice.
2. If you use the `attach` command, make sure that you use unique variable names. Refrain from common names such as Month, Location, and the like.
3. If you import multiple datasets, and only work with one dataset at a time, consider removing a data frame from the search path of R with the `detach` command.

In the remaining sections of this chapter, we assume that you did not type in the `attach(Squid)` command. If you did, type

```
> detach(Squid)
```

 Do Exercise 1 in Section 3.7. This is an exercise in using the `read.table` function and accessing variables using an epidemiological dataset.

3.2 Accessing Subsets of Data

In this section, we discuss how to access and extract components of the data frame `Squid`. The methods can be applied on a data frame that you created yourself by typing in data, as in Chapter 2.

The situation may arise in which you only want to work with, for example, the female data, data from a certain location, or data from the females of a certain location. To extract the subsets of data, we need to know how sex was coded. We could type in

```
> Squid$Sex
 [1] 2 2 2 2 2 2 2 2 2 2 2 2 2 2 2 2 2 2 2 2 2 2
[23] 2 1 2 2 2 2 2 2 2 2 2 2 2 2 2 2 2 2 2 2 2 2
[45] 2 2 2 1 2 2 2 2 2 2 2 2 2 1 2 1 1 1 1 2 1 1
[67] 1 1 1 1 1 1 1 1 1 2 1 1 1 1 1 1 1 1 1 1 1 1
```
                    < Cut to reduce space >

but this shows all values in the variable `Sex`. A better option is to use the `unique` command that shows how many unique values there are in this variable:

```
> unique(Squid$Sex)
[1] 2 1
```

The 1 stands for male, and the 2 for female. To access all the male data, use

```
> Sel <- Squid$Sex == 1
> SquidM <- Squid[Sel, ]
> SquidM
      Sample Year Month Location Sex    GSI
24        24    1     5        1   1 5.2970
48        48    1     5        3   1 4.2968
58        58    1     6        1   1 3.5008
60        60    1     6        1   1 3.2487
61        61    1     6        1   1 3.2304
```
< *Cut to reduce space* >

The first line creates a vector Sel that has the same length as the variable Sex, with values that are TRUE if Sex equals 1, and FALSE otherwise. Such a vector is also called a Boolean vector, and can be used to select rows, hence our name Sel. On the next line, we select the rows of Squid for which Sel equals TRUE, and we store the selected data in SquidM. Because we are selecting rows of Squid, we need to use the square brackets [ ], and, as we want rows, the vector Sel with Boolean values must go *before* the comma. It is also possible to do both lines in one command:

```
> SquidM <- Squid[Squid$Sex == 1, ]
> SquidM

      Sample Year Month Location Sex    GSI
24        24    1     5        1   1 5.2970
48        48    1     5        3   1 4.2968
58        58    1     6        1   1 3.5008
60        60    1     6        1   1 3.2487
61        61    1     6        1   1 3.2304
```
< *Cut to reduce space* >

The data for females are obtained by

```
> SquidF <- Squid[Squid$Sex == 2, ]
> SquidF
     Sample Year Month Location Sex     GSI
1         1    1     1        1   2 10.4432
2         2    1     1        3   2  9.8331
3         3    1     1        1   2  9.7356
4         4    1     1        1   2  9.3107
5         5    1     1        1   2  8.9926
```

< Cut to reduce space >

The process of selecting variable data (or data frames) conditional on the values of a second variable is called conditional selection. The `unique` command applied on `Squid$Location` shows that there are four locations coded as 1, 2, 3, and 4. To extract the data from location 1, 2, or 3 we can use the following statements that all give the same result (the | symbol stands for the Boolean "or" and the != for "not equal").

```
> Squid123 <- Squid[Squid$Location == 1 |
           Squid$Location == 2 | Squid$Location == 3, ]
> Squid123 <- Squid[Squid$Location != 4, ]
> Squid123 <- Squid[Squid$Location < 4, ]
> Squid123 <- Squid[Squid$Location <= 3, ]
> Squid123 <- Squid[Squid$Location >= 1 &
           Squid$Location <= 3, ]
```

You can choose any of these options. Next we use the "&," which is the Boolean "and" operator. Suppose we want to extract the male data from location 1. This means that the data have to be both from male squid *and* from location 1. The following code extracts data that comply with these conditions.

```
> SquidM.1 <- Squid[Squid$Sex == 1 &
              Squid$Location == 1,]
     Sample Year Month Location Sex    GSI
24       24    1     5        1   1 5.2970
58       58    1     6        1   1 3.5008
60       60    1     6        1   1 3.2487
61       61    1     6        1   1 3.2304
63       63    1     6        1   1 3.1848
```

< Cut to reduce space >

The data from males *and* from location 1 *or* 2 are given by

```
> SquidM.12 <- Squid[Squid$Sex == 1 &
         (Squid$Location == 1 | Squid$Location == 2), ]
```

Do *not* use the following command.

```
> SquidM <- Squid[Squid$Sex == 1, ]
> SquidM1 <- SquidM[Squid$Location == 1, ]
> SquidM1
     Sample Year Month Location Sex    GSI
24       24    1     5        1   1 5.2970
58       58    1     6        1   1 3.5008
```

```
60        60     1      6       1     1 3.2487
61        61     1      6       1     1 3.2304
62        62     1      5       3     1 3.2263
...
NA.1113        NA     NA     NA       NA   NA       NA
NA.1114        NA     NA     NA       NA   NA   NA
NA.1115        NA     NA     NA       NA   NA   NA
NA.1116        NA     NA     NA       NA   NA   NA
```

The first line extracts the male data and allocates it to SquidM, which is therefore of a smaller dimension (fewer rows) than Squid (assuming there are female squid in the data). On the next line, the Boolean vector Squid$Location == 1 is longer than the number of rows in SquidM, and R will add extra rows with NAs to SquidM. As a result, we get a data frame, SquidM1, that contains NAs. The problem is that we are trying to access elements of SquidM using a Boolean vector that has the same number of rows as Squid.

Don't panic if the output of a subselecting command shows the following message.

```
> Squid[Squid$Location == 1 & Squid$Year == 4 &
         Squid$Month == 1, ]
[1] Sample      Year        Month       Location   Sex
        GSI           fSex         fLocation
<0 rows> (or 0-length row.names)
```

This simply means that no measurements were taken at location 1 in month 1 of the fourth year.

3.2.1 Sorting the Data

In addition to extracting subsets of data, at times it is useful to rearrange the data. For the squid data, you may want to sort the GSI data from low to high values of the variable "month", even if only for a quick observation. The following code can be used.

```
> Ord1 <- order(Squid$Month)
> Squid[Ord1, ]
 Sample Year Month Location Sex      GSI
1      1    1     1         1   2 10.4432
2      2    1     1         3   2  9.8331
3      3    1     1         1   2  9.7356
4      4    1     1         1   2  9.3107
5      5    1     1         1   2  8.9926
```

< Cut to reduce space >

As we are manipulating the rows of Squid, we need to put Ord1 *before* the comma. We can also perform this exercise on only one variable, for instance the GSI. In this case, use

```
> Squid$GSI [Ord1]
 [1]   10.4432   9.8331   9.7356   9.3107   8.9926   8.7707
 [7]    8.2576   7.4045   7.2156   6.3882   6.0726   5.7757
[13]    1.2610   1.1997   0.8373   0.6716   0.5758   0.5518
[19]    0.4921   0.4808   0.3828   0.3289   0.2758   0.2506
```

< Cut to reduce space >

 Do Exercise 2 in Section 3.7. This is an exercise in using the read.- table function and accessing subsets from a data frame using a deep sea research dataset.

3.3 Combining Two Datasets with a Common Identifier

So far, we have seen examples in which all data points were stored in the same file. However, this may not always be the case. The authors of this book have been involved in various projects in which the data consisted of different types of measurements on the same animals. For example, in one project the measurements were made on approximately 1000 fish at different research institutes; one institute calculated morphometric measurements, another measured chemical variables, and yet another counted number of parasites. Each institute created its own spreadsheet containing the workgroup-specific variables. The crucial point was that, at each institute, researchers measured each fish, so all spreadsheets contained a column identifying the fish. Some fish were lost during the process or were unsuitable for certain procedures. Hence, the end result was a series of Excel spreadsheets, each with thousands of observations on 5–20 group-specific variables, but with a common identifier for the individual fish (case).

As a simple example of such a dataset, see the spreadsheets in Fig. 3.2. Imagine that the squid data were organised in this way, two different files or worksheets but with a common identifier. The task is now to merge the two datasets in such a way that data for sample *j* in the first dataset are put next to the data for sample *j* in the second dataset. For illustrative purposes, we have removed the fourth row from the second spreadsheet; just assume that someone forgot to type in Year, Month, Location, and Sex for observation 4. R has a useful tool for merging files, namely the *merge* function. It is run using the following code. The first two lines are used to read the two separate squid files:

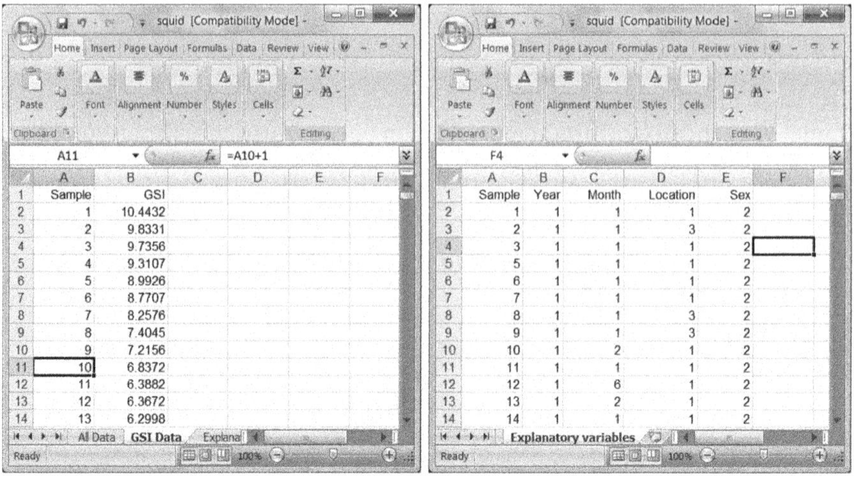

Fig. 3.2 GSI data with sample number (*left*) and the other variables with sample number (*right*). To illustrate the `merge` function, we deleted row number four in the right-hand spreadsheet

```
> setwd("C:/RBook/")
> Sq1 <- read.table(file = "squid1.txt",
                    header = TRUE)
> Sq2 <- read.table(file = "squid2.txt",
                    header = TRUE)
> SquidMerged <- merge(Sq1, Sq2, by = "Sample")
> SquidMerged
   Sample      GSI Year Month Location Sex
1       1  10.4432    1     1        1   2
2       2   9.8331    1     1        3   2
3       3   9.7356    1     1        1   2
4       5   8.9926    1     1        1   2
5       6   8.7707    1     1        1   2
6       7   8.2576    1     1        1   2
7       8   7.4045    1     1        3   2
8       9   7.2156    1     1        3   2
9      10   6.8372    1     2        1   2
10     11   6.3882    1     1        1   2
```

                        < *Cut to reduce space* >

The `merge` command takes as argument the two data frames `Sq1` and `Sq2` and combines the two datasets using as a common identifier the variable `Sample`. A useful option within the merge function is `all`. By default it is set to FALSE, which means that rows in which either `Sq1` or `Sq2` has missing

values are omitted. When set to TRUE, NAs are filled in if Sq1 has no data for a
sample that is present in Sq2, and vice versa. Using this option, we get

```
> SquidMerged <- merge(Sq1, Sq2, by = "Sample",
                       all = TRUE)
> SquidMerged
  Sample     GSI Year Month Location Sex
1       1 10.4432    1     1        1   2
2       2  9.8331    1     1        3   2
3       3  9.7356    1     1        1   2
4       4  9.3107   NA    NA       NA  NA
5       5  8.9926    1     1        1   2
6       6  8.7707    1     1        1   2
7       7  8.2576    1     1        1   2
8       8  7.4045    1     1        3   2
9       9  7.2156    1     1        3   2
10     10  6.8372    1     2        1   2
```

< Cut to reduce space >

Note the missing values for Year, Month, Location, and Sex for observation
(fish/case) 4. To avoid confusion, recall that we only removed the observations
from row four for illustrative purposes. Further options and examples are given
in the merge help file.

3.4 Exporting Data

In addition to the read.table command, R also has a write.table
command. With this function, you can export numerical information to an
ascii file. Suppose you extracted the male squid data, and you want to export it
to another software package, or send it to a colleague. The easiest way to do this
is to export the male squid data to an ascii file, then import it to the other
software package, or email it to your colleague. The following commands
extract the male data (in case you didn't type it in yet), and exports the data
to the file, MaleSquid.txt.

```
> SquidM <- Squid[Squid$Sex == 1, ]
> write.table(SquidM,
    file = "MaleSquid.txt",
    sep = " ", quote = FALSE, append = FALSE, na = "NA")
```

The first argument in the write.table function is the variable that you
want to export, and, obviously, you also need a file name. The sep = " "
ensures that the data are separated by a space, the quote = FALSE avoids

quotation marks around character strings (headings), na = "NA" allows you to specify how missing values are represented, and append = FALSE opens a new file. If it were set to TRUE, it would append the variable SquidM to the end of an existing file.

Let us illustrate some of these options. When we run the code above, the first six lines in the ascii file MaleSquid.txt are as follows.

```
Sample Year Month Location Sex GSI fLocation fSex
24 24 1 5 1 1 5.297 1 M
48 48 1 5 3 1 4.2968 3 M
58 58 1 6 1 1 3.5008 1 M
60 60 1 6 1 1 3.2487 1 M
61 61 1 6 1 1 3.2304 1 M
```
      < Cut to reduce space >

Hence, the elements are separated by a space. Note that we are missing the name of the first column. If you import these data into Excel, you may have to shift the first row one column to the right. We can change the sep and quote options:

```
> write.table(SquidM,
    file = "MaleSquid.txt",
    sep = ",", quote = TRUE, append = FALSE, na = "NA")
```

It gives the following output in the ascii file MalesSquid.txt.

```
"Sample","Year","Month","Location","Sex","GSI",
"fLocation","fSex"
"24",24,1,5,1,1,5.297,"1","M"
"48",48,1,5,3,1,4.2968,"3","M"
"58",58,1,6,1,1,3.5008,"1","M"
"60",60,1,6,1,1,3.2487,"1","M"
"61",61,1,6,1,1,3.2304,"1","M"
```

The fact that the headers extend over two lines is due to our text editor. The real differences are the comma separations and the quotation marks around categorical variables, and headers and labels. For some packages this is important. The append = TRUE option is useful if, for example, you have to apply linear regression on thousands of datasets and you would like to have all the numerical output in one file.

Do Exercise 3 in Section 3.7. This is an exercise in the write.table function using a deep sea research dataset.

3.5 Recoding Categorical Variables

In Section 3.1, we used the `str` function to give the following output for the Squid data frame.

```
> str(Squid)
'data.frame': 2644 obs. of 6 variables:
 $ Sample  : int 1 2 3 4 5 6 7 8 9 10 ...
 $ Year    : int 1 1 1 1 1 1 1 1 1 1 ...
 $ Month   : int 1 1 1 1 1 1 1 1 1 2 ...
 $ Location: int 1 3 1 1 1 1 1 3 3 1 ...
 $ Sex     : int 2 2 2 2 2 2 2 2 2 2 ...
 $ GSI     : num 10.44 9.83 9.74 9.31 8.99 ...
```

The variable `Location` is coded as 1, 2, 3, or 4, and `Sex` as 1 or 2. Such variables are categorical or nominal variables. In Excel, we could have coded sex as male and female. It is good programming practice to create new variables in the data frame that are recoded as nominal variables, for example:

```
> Squid$fLocation <- factor(Squid$Location)
> Squid$fSex <- factor(Squid$Sex)
```

These two commands create two new variables inside the data frame `Squid`, `fLocation` and `fSex`. We used the f in front of the variable name to remind us that these are nominal variables. In R, we can also call them factors, hence the f. Type

```
> Squid$fSex
  [1] 2 2 2 2 2 2 2 2 2 2 2 2 2 2 2 2 2
 [18] 2 2 2 2 2 2 1 2 2 2 2 2 2 2 2 2 2
 [35] 2 2 2 2 2 2 2 2 2 2 2 2 1 2 2 2
 ...
[2602] 1 2 1 1 2 1 1 1 2 1 2 1 1 2 1 1 2
[2619] 1 2 2 1 1 1 1 1 1 1 1 1 2 1 1 1 2
[2636] 1 2 1 2 1 2 1 1 1
Levels: 1 2
```

Note the extra line at the end. It tells us that `fSex` has two levels, 1 and 2. It is also possible to relabel these levels as "male" and "female", or, perhaps more efficiently, M and F:

```
> Squid$fSex <- factor(Squid$Sex, levels = c(1, 2),
  labels = c("M", "F"))
> Squid$fSex
  [1] F F F F F F F F F F F F F F F F F
 [18] F F F F F F M F F F F F F F F F F
 [35] F F F F F F F F F F F F M F F F
 ...
```

```
[2602] M F M M F M M M F M F M M F M M F
[2619] M F F M M M M M M M M M F M M M F
[2636] M F M F M F M M M
Levels: M F
```

Every 1 has been converted to an "M", and every 2 to an "F". You can now use fSex in functions such as lm or boxplot:

```
> boxplot(GSI ~ fSex, data = Squid) #Result not shown
> M1 <- lm(GSI ~ fSex + fLocation, data = Squid)
```

Another advantage of using predefined nominal variables in a linear regression function is that its output becomes much shorter. Although we do not show the output here, compare that of the following commands.

```
> summary(M1)
> M2 <- lm(GSI ~ factor(Sex) + factor(Location),
      data = Squid)
> summary(M2)
```

The estimated parameters are identical, but the second model needs more space on the screen (and on paper). This becomes a serious problem with second- and third-order interaction terms.

Instead of the command factor, you can also use as.factor. To convert a factor to a numerical vector, use as.numeric. This can be useful for making plots with different colours for males and females (if you have lost, for some reason, the original vector, Sex). See also Chapter 5.

The same can be done for fLocation:

```
> Squid$fLocation
 [1] 1 3 1 1 1 1 1 3 3 1 1 1 1 1 1 1 3
[18] 1 3 1 3 1 1 1 1 1 1 1 1 1 1 1 1 1
[35] 1 1 1 1 1 3 1 1 1 1 3 1 1 3 1 1 1
 ...
[2602] 1 1 1 1 1 1 1 1 1 1 1 1 1 1 1 1 1
[2619] 1 1 1 1 1 1 1 1 1 1 1 1 1 1 1 1 1
[2636] 1 1 1 1 1 1 1 1 1
Levels: 1 2 3 4
```

Note that this nominal variable has four levels. In this case, the levels are sorted from small to large. And this means that in a boxplot, the data from location 1 are next to location 2, 2 is next to 3, and so on. Sometimes it can be useful to change the order (e.g., in an xyplot function from the lattice package). This is done as follows.

```
> Squid$fLocation <- factor(Squid$Location,
                          levels = c(2, 3, 1, 4))
```

```
> Squid$fLocation
 [1] 1 3 1 1 1 1 1 3 3 1 1 1 1 1 1 1 3
[18] 1 3 1 3 1 1 1 1 1 1 1 1 1 1 1 1 1
[35] 1 1 1 1 1 3 1 1 1 1 3 1 1 3 1 1 1
...
[2602] 1 1 1 1 1 1 1 1 1 1 1 1 1 1 1 1 1
[2619] 1 1 1 1 1 1 1 1 1 1 1 1 1 1 1 1 1
[2636] 1 1 1 1 1 1 1 1 1
Levels: 2 3 1 4
```

The data values remain the same, but a command such as

```
> boxplot(GSI ~ fLocation, data = Squid)
```

produces a slightly different boxplot because the order of the levels is different. Relevelling is also useful for conducting a posthoc test in linear regression (Chapter 10 in Dalgaard, 2002).

We began this chapter with selecting the male data:

```
> SquidM <- Squid[Squid$Sex == 1, ]
```

We can also do this with fSex, but now we need:

```
> SquidM <- Squid[Squid$fSex == "1", ]
```

The quotation marks around the 1 are needed because fSex is a factor. The effect of defining new nominal variables can also be seen with the str command:

```
> Squid$fSex <- factor(Squid$Sex, labels = c("M", "F"))
> Squid$fLocation <- factor(Squid$Location)
> str(Squid)
'data.frame': 2644 obs. of 8 variables:
 $ Sample    : int 1 2 3 4 5 6 7 8 9 10 ...
 $ Year      : int 1 1 1 1 1 1 1 1 1 1 ...
 $ Month     : int 1 1 1 1 1 1 1 1 1 2 ...
 $ Location  : int 1 3 1 1 1 1 1 3 3 1 ...
 $ Sex       : int 2 2 2 2 2 2 2 2 2 2 ...
 $ GSI       : num 10.44 9.83 9.74 9.31 8.99 ...
 $ fSex      : Factor w/ 2 levels "M","F": 2 2 2 2 2 ...
 $ fLocation : Factor w/ 4 levels "1","2","3","4": 1 ...
```

Note that fSex and fLocation are now factors (categorical variables), and the levels are shown. Any function will now recognise them as factors, and there is no further need to use the factor command with these two variables.

Do Exercise 4 in Section 3.7. This is an exercise in use of the `factor` function using a deep sea research dataset.

3.6 Which R Functions Did We Learn?

Table 3.1 shows the R commands introduced in this chapter.

Table 3.1 R functions introduced in this chapter

Function	Purpose	Example
write.table	Write a variable to an ascii file	write.table(Z,file='' test.txt'')
order	Determine the order of the data	order(x)
merge	Merge two data frames	merge(x,y,by='' ID'')
attach	Make variables inside a data frame available	attach(MyData)
str	Shows the internal structure of an object	str(MyData)
factor	Defines a variable as a factor	factor(x)

3.7 Exercises

Exercise 1. Using the `read.table` function and accessing variables from a data frame with epidemiological data.

The file *BirdFlu.xls* contains the annual number of confirmed cases of human Avian Influenza A/(H5N1) for several countries reported to the World Health Organization (WHO). The data were taken from the WHO website (www.who.int/en/) and reproduced for educational purposes. Prepare the spreadsheet and import these data into R. If you are a non-Windows user, start with the file *BirdFlu.txt*. Note that you will need to adjust the column names and some of the country names.

Use the `names` and `str` command in R to view the data. Print the number of bird flu cases in 2003. What is the total number of bird flu cases in 2003 and in 2005? Which country has had the most cases? Which country has had the least bird flu deaths?

Using methods from Chapter 2, what is the total number of bird flu cases per country? What is the total number of cases per year?

Exercise 2. Using the `read.table` function and accessing subsets of a data frame with deep sea research data.

If you have not completed Exercise 6 in Chapter 2, do so and import the data from the *ISIT.xls* file.

In R, extract the data from station 1. How many observations were made at this station? What are the minimum, median, mean, and maximum sampled depth at station 1? What are the minimum, median, mean, and maximum sampled depth at station 2? At station 3?

Identify any stations with considerably fewer observations. Create a new data frame omitting these stations.

Extract the data from 2002. Extract the data from April (of all years). Extract the data that were measured at depths greater than 2000 meters (from all years and months). Show the data according to increasing depth values.

Show the data that were measured at depths greater than 2000 meters in April.

Exercise 3. Using the `write.table` function with deep sea research data.

In the final step of the previous exercise, data measured at depths greater than 2000 meters in April were extracted. Export these data to a new ascii file.

Exercise 4. Using the `factor` function and accessing subsets of a data frame with deep sea research data.

Stations 1 through 5 were sampled in April 2001, stations 6 through 11 in August 2001, stations 12 through 15 in March 2002, and stations 16 through 19 in October 2002. Create two new variables in R to identify the month and the year. Note that these are factors. Do this by adding the new variables inside the data frame.

Chapter 4
Simple Functions

In previous chapters, we demonstrated how to enter data; read data from a spreadsheet, ascii file, or a database; and extract subsets of data. In this chapter, we discuss applying some simple functions to the data, such as the mean or the mean of a single data subset. These are functions that may be useful; however, they are not the tools that will convince you to become an R user. Use them when it is convenient. Upon first reading of the book, you may skip this chapter.

4.1 The `tapply` Function

R provides functions for calculating the mean, length, standard deviation, minimum, maximum, variance, and any other function of a single variable, multiple variables, or on subsets of observations. For illustration, we use a vegetation dataset. Sikkink et al. (2007) analysed grassland data from a monitoring program conducted in two temperate communities, Yellowstone National Park and the National Bison Range, USA. The aim of the study was to determine whether the biodiversity of these bunchgrass communities changed over time, and, if so, whether the changes in biodiversity related to particular environmental factors. For our purposes we use only the Yellowstone National Park data. To quantify biodiversity, the researchers calculated species richness, defined as the number of *different* species per site. The study identified about 90 species. The data were measured in 8 transects, with each transect being assessed at intervals of 4–10 years, for a total of 58 observations.

The following code can be used to import the data and gain basic information on the variables.

```
> setwd("C:/RBook/")
> Veg <- read.table(file="Vegetation2.txt",
                     header= TRUE)
> names(Veg)
```

A.F. Zuur et al., *A Beginner's Guide to R*, Use R,
DOI 10.1007/978-0-387-93837-0_4, © Springer Science+Business Media, LLC 2009

```
[1] "TransectName"  "Samples"        "Transect"
[4] "Time"          "R"              "ROCK"
[7] "LITTER"        "ML"             "BARESOIL"
[10] "FallPrec"     "SprPrec"        "SumPrec"
[13] "WinPrec"      "FallTmax"       "SprTmax"
[16] "SumTmax"      "WinTmax"        "FallTmin"
[19] "SprTmin"      "SumTmin"        "WinTmin"
[22] "PCTSAND"      "PCTSILT"        "PCTOrgC"

> str(Veg)

'data.frame':   58 obs. of 24 variables:
$ TransectName: Factor w/ 58 levels ...
$ Samples     : int 1 2 3 4 5 6 7 8 9 10 ...
$ Transect    : int 1 1 1 1 1 1 1 2 2 2 ...
$ Time        : int 1958 1962 1967 1974 1981 1994...
$ R           : int 8 6 8 8 10 7 6 5 8 6 ...
$ ROCK        : num 27 26 30 18 23 26 39 25 24 21 ...
$ LITTER      : num 30 20 24 35 22 26 19 26 24 16 ...

            <Cut to reduce space>
```

The data are stored in the ascii file "Vegetation2.txt." Once the read.table function has been executed, we need to ensure that richness is indeed a numerical vector or integer. If, for some reason, R imports richness as a factor (e.g., because there is an alphanumerical in the column, or there are problems with the decimal separator), functions such as mean, sd, and the like will give an error message[1].

4.1.1 Calculating the Mean Per Transect

One of the first things we would like to know is whether the mean richness per transect differs. The code below calculates the mean richness, as well as mean richness for each transect (see Chapter 3 for selecting subsets of data):

```
> m <- mean(Veg$R)
> m1<- mean(Veg$R[Veg$Transect == 1])
> m2<- mean(Veg$R[Veg$Transect == 2])
> m3<- mean(Veg$R[Veg$Transect == 3])
```

---

[1] If you import data that are decimal comma separated with the default settings (i.e., with decimal point separation) or vice versa, R will see all variables as factors, and the strcommand will tell you so. Hence, to check that data were imported correctly, we recommend always using the str command on the imported data frame immediately after importing the data.

```
> m4<- mean(Veg$R[Veg$Transect == 4])
> m5<- mean(Veg$R[Veg$Transect == 5])
> m6<- mean(Veg$R[Veg$Transect == 6])
> m7<- mean(Veg$R[Veg$Transect == 7])
> m8<- mean(Veg$R[Veg$Transect == 8])
> c(m, m1, m2, m3, m4, m5, m6, m7, m8)

[1]  9.965517  7.571429  6.142857 10.375000 9.250000
[6] 12.375000 11.500000 10.500000 11.833333
```

The variable m contains the mean richness of all 8 transects, and m1 through m8 show the mean richness values per transect. Note that the mean command is applied to Veg $R, which is a vector of data. It is not a matrix; hence there is no need for a comma between the square brackets.

4.1.2 Calculating the Mean Per Transect More Efficiently

It is cumbersome to type eight commands to calculate the mean value per transect. The R function tapply performs the same operation as the code above (for m1 through m8), but with a single line of code:

```
> tapply(Veg$R, Veg$Transect, mean)
          1          2          3          4          5
   7.571429   6.142857  10.375000   9.250000  12.375000
          6          7          8
  11.500000  10.500000  11.833333
```

You can also run this code as

```
> tapply(X = Veg$R, INDEX = Veg$Transect, FUN = mean)
```

The tapply function splits the data of the first variable (R), based on the levels of the second variable (Transect). To each subgroup of data, it applies a function, in this case the mean, but we can also use the standard deviation (function sd), variance (function var), length (function length), and so on. The following lines of code calculate some of these functions for the vegetation data.

```
> Me <- tapply(Veg$R, Veg$Transect, mean)
> Sd <- tapply(Veg$R, Veg$Transect, sd)
> Le <- tapply(Veg$R, Veg$Transect, length)
> cbind(Me, Sd, Le)
          Me        Sd Le
1   7.571429 1.3972763   7
2   6.142857 0.8997354   7
3  10.375000 3.5831949   8
```

```
4  9.250000 2.3145502   8
5 12.375000 2.1339099   8
6 11.500000 2.2677868   8
7 10.500000 3.1464265   6
8 11.833333 2.7141604   6
```

Each row in the output gives the mean richness, standard deviation, and number of observations per transect. In a later chapter we discuss graphic tools to visualise these values.

4.2 The `sapply` and `lapply` Functions

To calculate the mean, minimum, maximum, standard deviation, and length of the full series, we still need to use mean (Veg$R), min (Veg$R), max (Veg$R), sd (Veg$R), and length (Veg$R). This is laborious if we wish to calculate the mean of a large number of variables such as all the numerical variables of the vegetation data. We specifically say "numerical" as one cannot calculate the mean of a factor. There are 20 numerical variables in the vegetation dataset, columns 5–25 of the data frame Veg. However, we do not need to type in the mean command 20 times. R provides other functions similar to the tapply to address this situation: the lapply and the sapply. The use of sapply and its output is given below:

```
> sapply(Veg[, 5:9], FUN= mean)

        R       ROCK     LITTER        ML   BARESOIL
 9.965517 20.991379 22.853448 1.086207 17.594828
```

To save space, we only present the results of the first five variables. It is important to realise that tapply calculates the mean (or any other function) for subsets of observations of a variable, whereas lapply and sapply calculate the mean (or any other function) of one or more variables, using *all* observations.

The word FUN stands for function, and must be written in capitals. Instead of the mean, you can use any other function as an argument for FUN, and you can write your own functions. So what is the difference between sapply and lapply? The major differences lie in the presentation of output, as can be seen in the following example.

```
> lapply(Veg[, 5:9], FUN= mean)
$R
[1] 9.965517
$ROCK
[1] 20.99138
```

```
$LITTER
[1] 22.85345
$ML
[1] 1.086207
$BARESOIL
[1] 17.59483
```

The output of lapply is presented as a list, whereas sapply gives it as a vector. The choice depends on the format in which you would like the output.

The variable that contains the data in lapply and sapply needs to be a data frame. This will not work:

```
> sapply(cbind(Veg$R, Veg$ROCK, Veg$LITTER, Veg$ML,
              Veg$BARESOIL), FUN = mean)
```

It will produce one long vector of data, because the output of the cbind command is not a data frame. It can easily be changed to a data frame:

```
> sapply(data.frame(cbind(Veg$R, Veg$ROCK, Veg$LITTER,
        Veg$ML, Veg$BARESOIL)), FUN = mean)
       X1         X2         X3         X4         X5
 9.965517  20.991379  22.853448   1.086207  17.594828
```

Note that we have lost the variable labels. To avoid this, make a proper data frame (Chapter 2) before running the sapply function. Alternatively, use the colnames function after combining the data with the cbind function.

 Do Exercise 1 in Section 4.6. This is an exercise in the use of the tapply, sapply, and lapply functions with a temperature dataset.

4.3 The summary Function

Another function that gives basic information on variables is the summary command. The argument can be a variable, the output from a cbind command, or a data frame. It is run by the following commands.

```
> Z <-cbind(Veg$R, Veg$ROCK, Veg$LITTER)
> colnames(Z) <- c("R", "ROCK", "LITTER")
> summary(Z)
```

```
         R                    ROCK              LITTER
Min.    : 5.000     Min.     : 0.00   Min.      : 5.00
1st Qu. : 8.000     1st Qu. : 7.25   1st Qu. :17.00
Median  :10.000     Median  :18.50   Median  :23.00
Mean    : 9.966     Mean     :20.99   Mean      :22.85
3rd Qu. :12.000     3rd Qu. :27.00   3rd Qu. :28.75
Max.    :18.000     Max.     :59.00   Max.      :51.00
```

The summary command gives the minimum, first quartile, median, mean, third quartile, and maximum value of the variable. An alternative R code gives the same result:

```
> summary(Veg[ , c("R","ROCK","LITTER")])
```

or

```
> summary(Veg[ , c(5, 6, 7)])
```

Output is not presented here.

4.4 The `table` Function

In Exercises 1 and 7 in Section 2.4, we introduced the deer data from Vicente et al. (2006). The data were from multiple farms, months, years, and sexes. One of the aims of the study was to find a relationship between length of the animal and the number of *E. cervi* parasites. It may be the case that this relationship changes with respect to sex, year, month, farm, or even year and month. To test this, one needs to include interactions in the statistical models. However, problems may be encountered if there are no sampled females in some years, or if some farms were not sampled in every year. The table function can be used to learn how many animals per farm were sampled, as well as the number of observations per sex and year. The following code imports the data, and shows the results.

```
> setwd("c:/RBook/")
> Deer <- read.table(file="Deer.txt", header= TRUE)
> names (Deer)
[1] "Farm"    "Month"    "Year"     "Sex"      "clas1_4"
[6] "LCT"    "KFI"      "Ecervi"   "Tb"

> str(Deer)

[1] "Farm"    "Month"    "Year"     "Sex"      "clas1_4"
[6] "LCT"    "KFI"      "Ecervi"   "Tb"
```

```
> str(Deer)

'data.frame': 1182 obs. of 9 variables:
 $ Farm     : Factor w/ 27 levels"AL","AU","BA",..: 1...
 $ Month    : int 10 10 10 10 10 10 10 10 10 10 ...
 $ Year     : int 0 0 0 0 0 0 0 0 0 0 ...
 $ Sex      : int 1 1 1 1 1 1 1 1 1 1 ...
 $ clas1_4  : int 4 4 3 4 4 4 4 4 4 4 ...
 $ LCT      : num 191 180 192 196 204 190 196 200 19 ...
 $ KFI      : num 20.4 16.4 15.9 17.3 NA ...
 $ Ecervi   : num 0 0 2.38 0 0 0 1.21 0 0.8 0 ...
 $ Tb       : int 0 0 0 0 NA 0 NA 1 0 0 ...
```

Farm has been coded as AL, AU, and so on, and is automatically imported as a factor. The other variables are all vectors of numerical or integer values. The number of observations per farm is obtained by

```
> table(Deer$Farm)
 AL   AU   BA   BE   CB  CRC   HB  LCV   LN  MAN   MB
 15   37   98   19   93   16   35    2   34   76   41
 MO   NC   NV   PA   PN   QM   RF   RÑ   RO  SAL  SAU
278   32   35   11   45   75   34   25   44    1    3
 SE   TI   TN VISO   VY
 26   21   31   15   40
```

At one farm, 278 animals were sampled and, at others, only one. This dataset typically requires a mixed effects modelling[2] approach in which "farm" is used as a random effect (see Zuur et al., 2009). This method can cope with unbalanced designs. However, the inclusion of a sex/year interaction term[3] in such models for these data will give an error message. This is because both sexes were not measured in every year, as can be seen from the following contingency table. (The labels 0, 1, 2, 3, 4, 5, and 99 in the horizontal direction refer to 2000, 2001, 2002, 2003, 2004, 2005, and 1999, respectively. In the vertical direction 1 and 2 indicate sex).

```
> table(Deer$Sex, Deer$Year)

     0   1   2   3   4   5  99
1  115  85 154  75  78  34  21
2   76  40 197 123  60  35   0
```

In 1999, animals of only one sex were measured. We recommend always using the `table` command before including interactions between two categorical variables in regression type models.

---

[2] A mixed effects model is an extension of linear regression.

[3] A sex/year interaction term allows the effect of sex to differ over the years.

Do Exercise 2 in Section 4.6. This is an exercise in using the `table` function with a temperature dataset.

4.5 Which R Functions Did We Learn?

Table 4.1 shows the R functions that were introduced in this chapter.

Table 4.1 R functions introduced in this chapter

Function	Purpose	Example
tapply	Apply FUN on y for each level of x	tapply (y, x, FUN = mean)
sapply	Apply FUN on each variable in y	sapply (y, FUN = mean)
lapply	Apply FUN on each variable in y	tapply (y, FUN = mean)
sd	Calculate the standard deviation of y	sd (y)
length	Determine the length of y.	length (y)
summary	Calculate general information	summary (y)
table	Calculate a contingency table	table (x, y)

4.6 Exercises

Exercise 1. The use of the `tapply`, `sapply`, and `lapply` functions to calculate mean temperature per month.

The file *temperature.xls* contains temperature observations made at 31 locations along the Dutch coastline. The data were collected and provided by the Dutch institute RIKZ (under the monitoring program MWTL; Monitoring Waterstaatkundige Toestand des Lands). Sampling began in 1990, and the final measurements in the spreadsheet were taken in December 2005, a period of 16 years. Sampling frequency was 0–4 times per month, depending on the season.

Calculate a one-time series of monthly averages using data from all stations. The end result should be a variable of dimension 16 × 12. Also calculate the standard deviation and number of observations per month.

Exercise 2. The use of the `table` function for the temperature data.

Using the data in Exercise 1, determine the number of observations per station. How many observations were made per year? How many observations were made at each station per year?

Chapter 5
An Introduction to Basic Plotting Tools

We have demonstrated the use of R tools for importing data, manipulating data, extracting subsets of data, and making simple calculations, such as mean, variance, standard deviation, and the like. In this chapter, we introduce basic graph plotting tools. If you are interested in only simple graphing, this chapter will suffice; however, to construct more sophisticated graphs, or to add more complicated embellishments such as tick marks, or specialized fonts and font sizes, to basic graphs, you will need the more advanced plotting techniques presented in Chapters 7 and 8.

A discussion of elementary plotting tools may seem out of place at this stage, rather than being included in the sections on graphing beginning with Chapter 7. However, when teaching the material presented in this book, we became aware that, after discussing the relatively pedestrian material of the first four sections, the course participants were eagerly awaiting the lively, more visual, and easier, plotting tools. Therefore, we present a first encounter with graphing here, which allows the presentation of the more complex subjects in the next chapter with the aid of active tools such as the `plot` function.

5.1 The `plot` Function

This section uses the vegetation data introduced in Chapter 4. Recall that these are grassland data from a monitoring program conducted in two temperate communities in Yellowstone National Park and the National Bison Range, USA. To quantify biodiversity, species richness was calculated. In a statistical analysis, we may want to model richness as a function of BARESOIL (or any of the other soil and climate variables). Suppose we want to make a plot of species richness versus the substrate variable "exposed soil," denoted by BARESOIL. The R commands to create such a graph is

```
> setwd("c:/RBook/")
> Veg <- read.table(file = "Vegetation2.txt",
                    header = TRUE)
> plot(Veg$BARESOIL, Veg$R)
```

A.F. Zuur et al., *A Beginner's Guide to R*, Use R,
DOI 10.1007/978-0-387-93837-0_5, © Springer Science+Business Media, LLC 2009

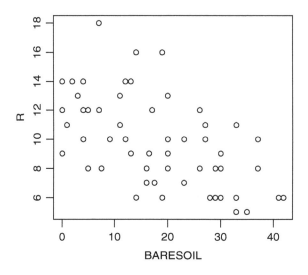

Fig. 5.1 Graph showing a scatterplot of BARESOIL versus species richness for the vegetation data. To import a graph from R into Microsoft Word, right-click on the graph in R to copy and paste it into Word, or save it as a metafile (recommended, as it produces a good quality graph), bitmap, or postscript file. This will also work with non-Windows operating systems. Importing the latter two formats into Word is more complicated. If the R console window (Fig. 1.5) is maximised, you may not see the panel with the graph. To access it, press Control-Tab, or minimise the R console

The resulting graph is presented in Fig. 5.1. The first argument of the plot command is displayed on the horizontal axis with the second argument along the vertical axis. Richness, in this case, is the response, or dependent, variable, and BARESOIL is the explanatory, or independent, variable. It is conventional to plot the response variable along the vertical axis and the explanatory variable along the horizontal axis. Be aware that for some statistical functions in R, you must specify the response variable first, followed by the explanatory variables. You would not be the first to accidentally type

```
> plot(Veg$R, Veg$BARESOIL)
```

and discover that the order should have been reversed. Alternatively, you can use

```
> plot(x = Veg$BARESOIL, y = Veg$R)
```

to avoid confusion over which variables will be plotted on the *x*-axis (horizontal) and which on the *y*-axis (vertical).

The plot function does have a data argument, but we cannot use

```
> plot(BARESOIL, R, data = Veg)
Error in plot(BARESOIL, R, data = Veg): object
"BARESOIL" not found
```

This is unfortunate, and forces us to use Veg$ in the command (recall from Chapter 2 that we do not use the attach command). This is one of the reasons (if not the only one) that we chose the variable name Veg instead of the longer, Vegetation.It is also possible to use:

```
> plot(R ~ BARESOIL, data = Veg)
```

This does produce a graph (not shown here), but our objection against this notation is that in some functions, the R ~ BARESOIL notation is used to tell R that richness is modelled *as a function of* baresoil. Athough that may be the case here, not every scatterplot involves variables that have a cause–effect relationship.

The most common modifications to any graph are adding a title and *x*- and *y*-labels and setting the *x*- and *y*-limits, as with the graph in Fig. 5.1. This is accomplished by extending the plot command:

```
> plot(x = Veg$BARESOIL, y = Veg$R,
     xlab = "Exposed soil",
     ylab = "Species richness", main = "Scatter plot",
     xlim = c(0, 45), ylim = c(4, 19))
```

The resulting graph is presented in Fig. 5.2A. The four panels (A–D) were imported into the text document (Microsoft Word) by using the text editor to construct a 2-by-2 table with no borders. (Using R to create a graph with multiple panels is demonstrated in Chapter 8.)

The order in which xlab, ylab , main, xlim, and ylim are entered is irrelevant, but they must be in lowercase letters. The xlab and ylab options are used for labels and the main option for the title. The xlim and ylim options are used to specify the lower and upper limits on the axes. You can also use

```
xlim = c(min(Veg$BARESOIL), max(Veg$BARESOIL))
```

within the plot command, but, if there are missing values in the plotted variable, you should extend the min and max functions with the na.rm = TRUE option. This produces

```
xlim = c(min(Veg$BARESOIL, na.rm = TRUE),
         max(Veg$BARESOIL, na.rm = TRUE))
```

In Chapters 7 and 8, we demonstrate changing the style and size of the font used for the labels and title, and adding symbols, such as, ^{0}C, and so on.

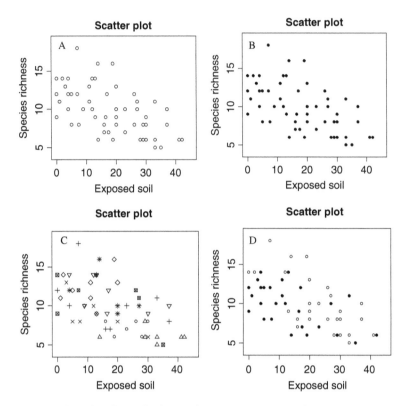

Fig. 5.2 Examples of various plotting options. **A**: Scatterplot of species richness versus exposed soil (BARESOIL). **B**: The same scatterplot as in A, with observations plotted using a *filled circle* (*or dot*). **C**: The same scatterplot as in A, with observations from each transect indicated by a different symbol. **D**: The same scatterplot as in A, but with observations measured before 1975 plotted as *open circles* and those after 1975 as *filled circles*

5.2 Symbols, Colours, and Sizes

During our courses, the most frequently asked questions concerning graphs are whether (1) the plotting symbols can be changed, (2) different colours can be used, and (3) the size of the plotting symbols can be varied conditional on the values of another variable. We discuss these three options in this section and leave the more sophisticated modifications such as altering tick marks, and adding subscripts and superscripts, among other things, for Chapters 7 and 8.

5.2.1 Changing Plotting Characters

By default, the plot function uses open circles (open dots) as plotting characters, but characters can be selected from about 20 additional symbols. The plotting character is specified with the pch option in the plot function; its default value is 1 (which is the open dot or circle). Figure 5.3 shows the symbols that can be obtained with the different values of pch. For solid points the command is pch = 16. As an example, the following code produces Fig. 5.2B, in which we replaced the open dots with filled dots.

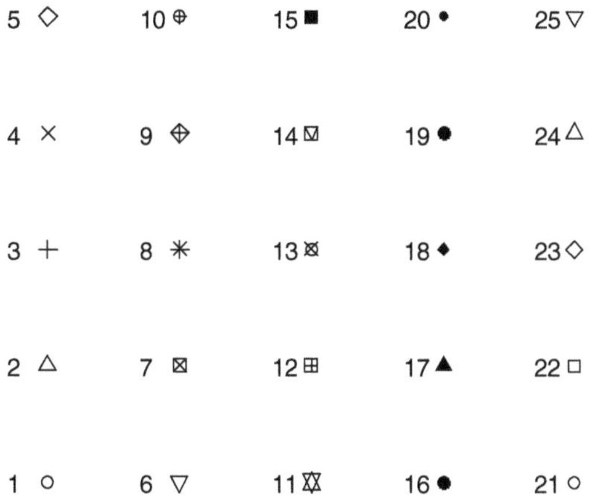

Fig. 5.3 Symbols that can be obtained with the pch option in the plot function. The number left of a symbol is the pch value (e.g., pch = 16 gives ●)

5 ◇	10 ⊕	15 ■	20 •	25 ▽
4 ✕	9 ⊕	14 ⊠	19 ●	24 △
3 +	8 ✳	13 ⊠	18 ◆	23 ◇
2 △	7 ⊠	12 ⊞	17 ▲	22 □
1 ○	6 ▽	11 ⊠	16 ●	21 ○

```
> plot(x = Veg$BARESOIL, y = Veg$R,
      xlab = "Exposed soil",
      ylab = "Species richness", main = "Scatter plot",
      xlim = c(0, 45), ylim = c(4, 19), pch = 16)
```

In Fig. 5.2A, B, all observations are represented by the same plotting symbol (the open circles in panel A were obtained with the default, pch = 1, and the closed circles in panel B with pch = 16).

The grassland data were measured over the course of several years in eight transects. It would be helpful to add this information to the graph in Fig. 5.2A. At this point, the flexibility of R begins to emerge. Suppose you want to use a different symbol for observations from each transect. To do this, use a numerical vector that has the same length as BARESOIL and richness R and contains the value 1 for all observations from transect 1, the value 2 for all observations from transect 2, and so on. Of course it is not necessary to use 1, 2, and so on. The values can be any valid pch number (Fig. 5.3). You only need to ensure that, in the new numerical vector, the values for observations within a single transect are the same and are different from those of the other transects. In this case you are lucky; the variable Transect is already coded with numbers 1 through 8 designating the eight transects. To see this, type

```
> Veg$Transect
 [1] 1 1 1 1 1 1 1 1 2 2 2 2 2 2 2 3 3 3 3 3 3 3 3
[23] 4 4 4 4 4 4 4 4 5 5 5 5 5 5 5 5 6 6 6 6 6 6
[45] 6 6 7 7 7 7 7 7 8 8 8 8 8 8
```

Thus, there is no need to create a new vector; you can use the variable Transect (this will not work if Transect is defined as a factor, see below):

```
> plot(x = Veg$BARESOIL, y = Veg$R,
    xlab = "Exposed soil", ylab = "Species richness",
    main = "Scatter plot", xlim = c(0, 45),
    ylim = c(4, 19), pch = Veg$Transect)
```

The resulting graph is presented in Fig. 5.2C. It shows no clear transect effect. It is not a good graph, as there is too much information, but you have learned the basic process.

There are three potential problems with the pch = Transect approach:

1. If Transect had been coded as 0, 1, 2, and so on, the transect for which pch = 0 would not have been plotted.
2. If the variable Transect did not have the same length as BARESOIL and richness R, assume it was shorter; R would have repeated (iterated) the first elements in the vector used for the pch option, which would obviously produce a misleading plot. In our example, we do not have this problem, as BARESOIL, richness, and transect have the same length.
3. In Chapter 2, we recommended that categorical (or nominal) variables be defined as such in the data frame using the factor command. If you select a nominal variable as the argument for pch, R will give an error message. This error message is illustrated below:

```
> Veg$fTransect <- factor(Veg$Transect)
> plot(x = Veg$BARESOIL, y = Veg$R,
      xlab = "Exposed soil",
      ylab = "Species richness", main = "Scatter plot",
      xlim = c(0, 45), ylim = c(4, 19),
      pch = Veg$fTransect)
Error in plot.xy(xy, type, ...): invalid plotting symbol
```

On the first line of the R code above, we defined fTransect as a nominal variable inside the Veg data frame, and went on to use it as argument for the pch option. As you can see, R will not accept a factor as pch argument; it must be a numerical vector.

5.2.1.1 Use of a Vector for pch

The use of a vector for pch (and for the col and cex options discussed later) can be confusing.

The vegetation data were measured in 1958, 1962, 1967, 1974, 1981, 1989, 1994, and 2002. We arbitrarily selected an open circle to represent observations measured from 1958 to 1974 and a filled circle for those made after 1974. Obviously, the option pch = Veg$Time is out of the question, as it tries to

use eight different symbols, and, also, the pch value 1958 (or of any year) does
not exist. We must create a new numerical vector of the same length as Veg
$Time, using the value 1 when Time is 1958, 1962, 1967, and 1974 and 16 for
the more recent years. The values 1 and 16 were chosen because we like open
and filled circles as they show a greater contrast than other combinations. Here
is the R code (you can also do this in one line with the ifelse command):

```
> Veg$Time2 <- Veg$Time
> Veg$Time2 [Veg$Time <= 1974] <- 1
> Veg$Time2 [Veg$Time > 1974] <- 16
> Veg$Time2
```

```
 [1]  1 1 1  1 16 16 16  1 1 1  1  6 16 16  1
[16]  1 1 1 16 16 16 16  1 1 1  1 16 16 16 16
[31]  1 1 1  1 16 16 16 16 1 1  1  1 16 16 16
[46] 16 1 1  1 16 16 16  1 1 1 16 16 16
```

The first command creates a new numerical vector of the same length as Veg
$Time, and the following two commands allocate the values 1 and 16 to the
proper places. The rest of the R code is easy; simply use Veg $Time2 as the pch
option. The resulting graph is presented in Fig. 5.2D:

```
> plot(x = Veg$BARESOIL, y = Veg$R,
      xlab = "Exposed soil",
      ylab = "Species richness", main = "Scatter plot",
      xlim = c(0, 45), ylim = c(4, 19),
      pch = Veg$Time2)
```

In the text above, we mentioned that you should not use pch =Veg $Time
as Time contains values that are not valid pch commands. The use of Veg
$Time will result in

```
> plot(x = Veg$BARESOIL, y = Veg$R,
      xlab = "Exposed soil",
      ylab = "Species richness", main = "Scatter plot",
      xlim = c(0, 45), ylim = c(4, 19),
      pch = Veg$Time)
```

```
There were 50 or more warnings (use warnings() to see
the first 50)
```

```
> warnings()
```

```
Warning messages:
1: In plot.xy(xy, type, ...) : unimplemented pch value
' 1958'
```

```
2: In plot.xy(xy, type, ...) : unimplemented pch value
'1962'
3: In plot.xy(xy, type, ...) : unimplemented pch value
'1967'
4: In plot.xy(xy, type, ...) : unimplemented pch value
'1974'
5: In plot.xy(xy, type, ...) : unimplemented pch value
'1981'
....
```

We typed warnings () as instructed by R. The warning message speaks for itself.

To learn more about the pch option, look at the help file of the function points, obtained with the ? points command.

5.2.2 Changing the Colour of Plotting Symbols

The plotting option for changing colours is useful for graphics presented on a screen or in a report, but is less so for scientific publications, as these are most often printed in black and white. We recommend that you read Section 5.2.1 before reading this section, as the procedure for colour is the same as that for symbols.

To replace the black dots in Fig. 5.2 with red, use

```
> plot(x = Veg$BARESOIL, y = Veg$R,
    xlab = "Exposed soil",
    ylab = "Species richness", main = "Scatter plot",
    xlim = c(0, 45), ylim = c(4, 19),
    col = 2)
```

For green, use col = 3. Run the following code to see the other available colours.

```
> x <- 1:8
> plot(x, col = x)
```

We do not present the results of the two commands as this book is without colour pages. In fact, there are considerably more colours available in R than these eight. Open the par help file with the ? par command, and read the "Color Specification" section near the end. It directs you to the function colors (or colours), where, apparently, you can choose from hundreds.

5.2.2.1 Use of a Vector for `col`

You can also use a vector for the `col` option in the `plot` function. Suppose you want to plot the observations from 1958 to 1974 as black filled squares and the observations from 1981 to 2002 as red filled circles (shown here as light grey). In the previous section, you learned how to create filled squares and circles using the variable `Time2` with values 15 (square) and 16 (circle). Using two colours is based on similar R code. First, create a new variable of the same length as `BARESOIL` and richness `R`, which can be called `Col2` . For those observations from1958 to 1974, `Col2` takes the value 1 (= black) and, for the following years, 2 (= red). The R code is

```
> Veg$Time2 <- Veg$Time
> Veg$Time2 [Veg$Time <= 1974] <- 15
> Veg$Time2 [Veg$Time > 1974] <- 16
> Veg$Col2 <- Veg$Time
> Veg$Col2 [Veg$Time <= 1974] <- 1
> Veg$Col2 [Veg$Time > 1974] <- 2
> plot(x = Veg$BARESOIL, y = Veg$R,
      xlab = "Exposed soil",
      ylab = "Species richness", main = "Scatter plot",
      xlim = c(0, 45), ylim = c(4, 19),
      pch = Veg$Time2, col = Veg$Col2)
```

The resulting graph is presented in Fig. 5.4A. The problems that were outlined for the `pch` option also apply to the `col` option. If you use `col = 0`, the observations will not appear in a graph having a white background; the vector with values for the colours should have the same length as `BARESOIL` and richness `R`; and you must use values that are linked to a colour in R.

Before expending a great deal of effort on producing colourful graphs, it may be worth considering that, in some populations, 8% of the male population is colourblind!

5.2.3 Altering the Size of Plotting Symbols

The size of the plotting symbols can be changed with the `cex` option, and again, this can be added as an argument to the `plot` command. The default value for `cex` is 1. Adding `cex = 1.5` to the plot command produces a graph in which all points are 1.5 times the default size:

```
> plot(x = Veg$BARESOIL, y = Veg$R,
      xlab = "Exposed soil", ylab = "Species richness",
      main = "Scatter plot",
      xlim = c(0, 45), ylim = c(4, 19),
      pch = 16, cex = 1.5)
```

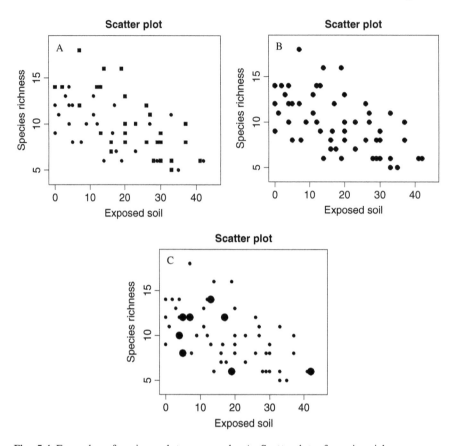

Fig. 5.4 Examples of various plot commands. **A**: Scatterplot of species richness versus BARESOIL. Observations from 1958 to 1974 are represented as filled squares in black and observations from 1981 to 2002 as *filled circles* in red. Colours were converted to greyscale in the printing process. **B**: The same scatterplot as in Fig. 5.2A, with all observations represented as black filled dots 1.5 times the size of the dots in Fig. 5.2A. **C**: The same scatterplot as in Fig. 5.2A with observations from 2002 represented by dots twice those of Fig. 5.2A

We used filled circles. The resulting graph is presented in Fig. 5.4B.

5.2.3.1 Use of a Vector for cex

As with the pch and col options, we demonstrate the use of a vector as the argument of the cex option. Suppose you want to plot BARESOIL against species richness using a large filled dot for observations made in 2002 and a smaller filled dot for all other observations. Begin by creating a new vector with values of 2 for observations made in 2002 and 1 for those from all other years. The values 1 and 2 are good starting points for finding, through trial and error, the optimal size difference. Try 3 and 1, 1.5 and 1, or 2 and 0.5, and so on, and decide which looks best.

```
> Veg$Cex2 <- Veg$Time
> Veg$Cex2[Veg$Time == 2002] <- 2
> Veg$Cex2[Veg$Time != 2002] <- 1
```

Using the vector Cex2 , our code can easily be adjusted:

```
> plot(x = Veg$BARESOIL, y = Veg$R,
       xlab = "Exposed soil", ylab = "Species richness",
       main = "Scatter plot",
       xlim = c(0, 45), ylim = c(4, 19),
       pch = 16, cex = Veg$Cex2)
```

The resulting graph is presented in Fig. 5.4C. Altering the symbol size can also be accomplished by using cex = 1.5 * Veg$Cex2 or cex = Veg $Cex2 /2.

5.3 Adding a Smoothing Line

It is difficult to see a pattern in Fig. 5.1. The information that you want to impart to the viewer will become clearer if you add a smoothing curve[1] to aid in visualising the relationship between species richness and BARESOIL. The underlying principle of smoothing is not dealt with in this book, and we refer the interested reader to Hastie and Tibshirani (1990), Wood (2006), or Zuur et al. (2007).

The following code redraws the plot, applies the smoothing method, and superimposes the fitted smoothing curve over the plot, through the use of the lines command.

```
> plot(x = Veg$BARESOIL, y = Veg$R,
     xlab = "Exposed soil", ylab = "Species richness",
     main = "Scatter plot", xlim = c(0, 45),
     ylim = c(4, 19))
> M.Loess <- loess(R ~ BARESOIL, data = Veg)
> Fit <- fitted(M.Loess)
> lines(Veg$BARESOIL, Fit)
```

The resulting graph is presented in Fig. 5.5A. The command

---

[1] A smoothing curve is a line that follows the shape of the data. For our purposes, it is sufficient to know that a smoothing curve serves to capture the important patterns in, or features of, the data.

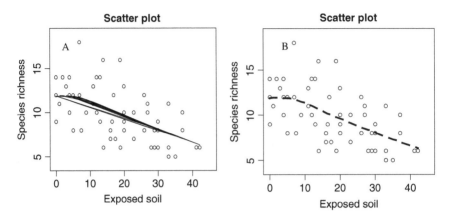

Fig. 5.5 A: The same scatterplot as in Fig. 5.2A, with a smoothing curve. Problems occur with the lines command because BARESOIL is not sorted from low to high. **B**: The same scatterplot as in Fig. 5.2A, but with a properly drawn smoothing curve

```
> M.Loess <- loess(R ~ BARESOIL, data = Veg)
```

is the step that applies the smoothing method, and its output is stored in the object M.Loess . To see what it comprises, type:

```
> M.Loess
```

```
Call:
loess(formula = R ~ BARESOIL, data = Veg)
Number of Observations: 58
Equivalent Number of Parameters: 4.53
Residual Standard Error: 2.63
```

That is not very useful. M.Loess contains a great deal of information which can be extracted through the use of special functions. Knowing the proper functions and how to apply them brings us into the realm of statistics; the interested reader is referred to the help files of resid, summary, or fitted (and, obviously, loess).

The notation R ~BARESOIL means that the species richness R is modelled as a function of BARESOIL. The loess function allows for various options, such as the amount of smoothing, which is not discussed here as it brings us even further into statistical territory. As long as we do not impose further specifications on the loess function, R will use the default settings, which are perfect for our purpose: drawing a smoothing curve.

The output from the loess function, M.Loess , is used as input into the function, fitted. As the name suggests, this function extracts the fitted values, and we allocate it to the variable Fit. The last command,

```
> lines(Veg$BARESOIL, Fit)
```

superimposes a line onto the plot that captures the main pattern in the data and transfers it onto the graph. The first argument goes along the *x*-axis and the second along the *y*-axis. The resulting plot is given in Fig. 5.5A. However, the smoothed curve is not what we expected, as the lines form a spaghetti pattern (multiple lines). This is because the `lines` command connects points that are sequential in the first argument.

There are two options for solving this problem. We can sort BARESOIL from small to high values and permute the second argument in the `lines` command accordingly, or, alternatively, we can determine the order of the values in BARESOIL, and rearrange the values of both vectors in the `lines` command. The second option is used below, and the results are given in Fig. 5.5B. Here is the R code.

```
> plot(x =  Veg$BARESOIL, y = Veg$R,
     xlab = "Exposed soil",
     ylab = "Species richness", main = "Scatter plot",
     xlim = c(0, 45), ylim = c(4, 19))
> M.Loess <- loess(R ~ BARESOIL, data = Veg)
> Fit <- fitted(M.Loess)
> Ord1 <- order(Veg$BARESOIL)
> lines(Veg$BARESOIL[Ord1], Fit[Ord1],
         lwd = 3, lty = 2)
```

The `order` command determines the order of the elements in BARESOIL, and allows rearranging of the values from low to high in the `lines` command. This is a little trick that you only need to see once, and you will use it many times thereafter. We also added two more options to the `lines` command, `lwd` and `lty`, indicating line width and line type. These are further discussed in Chapter 7, but to see their effect, change the numbers and note the change in the graph. Within the `lines` command, the `col` option can also be used to change the colour, but obviously the `pch` option will have no effect.

The smoothing function seems to indicate that there is a negative effect of BARESOIL on species richness.

5.4 Which R Functions Did We Learn?

Table 5.1 shows the R functions that were introduced in this chapter.

5.5 Exercises

Exercise 1. Use of the `plot` function using terrestrial ecology data. In Chapter 16 of Zuur et al. (2009), a study is presented analysing numbers of amphibians

Table 5.1 R functions introduced in this chapter

Function	Purpose	Example
plot	Plots *y* versus *x*	plot(y, x, xlab="X label", xlim=c(0, 1), pch=1, main="Main", ylim=c(0, 2), ylab="Y label", col=1)
lines	Adds lines to an existing graph	lines(x, y, lwd=3, lty=1, col=1)
order	Determines the order of the data	order(x)
loess	Applies LOESS smoothing	M<-loess(y~x)
fitted	Obtains fitted values	fitted(M)

killed along a road in Portugal using generalised additive mixed modelling techniques. In this exercise, we use the plot command to visualise a segment of the data. Open the file *Amphibian_road_Kills.xls*, prepare a spreadsheet, and import the data into R.

The variable, TOT_N, is the number of dead animals at a sampling site, OLIVE is the number of olive groves at a sampling site, and D Park is the distance from each sampling point to the nearby natural park. Create a plot of TOT_N versus D_park. Use appropriate labels. Add a smoothing curve. Make the same plot again, but use points that are proportional to the value of OLIVE (this may show whether there is an OLIVE effect).

Chapter 6
Loops and Functions

When reading this book for the first time, you may skip this chapter, as building functions[1] and programming loops[2] are probably not among the first R procedures you want to learn, unless these subjects are your prime interests. In general, people perceive these techniques as difficult, hence the asterisk in the chapter title. Once mastered, however, these tools can save enormous amounts of time, especially when executing a large number of similar commands.

6.1 Introduction to Loops

One of R's more convenient features is the provision for easily making your own functions. Functions are useful in a variety of scenarios. For example, suppose you are working with a large number of multivariate datasets, and for each of them you want to calculate a diversity index. There are many diversity indices, and new ones appear regularly in the literature. If you are lucky, the formula for your chosen diversity index has already been programmed by someone else, and, if you are very lucky, it is available in one of the popular packages, the software code is well documented, fully tested, and bug free. But if you cannot find software code for the chosen diversity index, it is time to program it yourself!

If you are likely to use a set of calculations more than once, you would be well advised to present the code in such a way that it can be reused with minimal typing. Quite often, this brings you into the world of functions and loops (and conditional statements such as the if command).

The example presented below uses a dataset on owls to produce a large number of graphs. The method involved is repetitive and time consuming, and a procedure that will do the hard work will be invaluable.

---

[1] A function is a collection of codes that performs a specific task.
[2] A loop allows the program to repeatedly execute commands. It does this by iteration (iteration is synonymous with repetition).

A.F. Zuur et al., *A Beginner's Guide to R*, Use R,
DOI 10.1007/978-0-387-93837-0_6, © Springer Science+Business Media, LLC 2009

Developing this procedure requires programming and some logical thinking. You will need to work like an architect who draws up a detailed plan for building a house. You should definitely not begin entering code for a function or loop until you have an overall design.

You also must consider how foolproof your function needs to be. Do you intend to use it only once? Should it work next year on a similar dataset (when you have forgotten most settings and choices in your function)? Will you share it with colleagues?

Functions often go hand in hand with loops, as they both help to automate commands.

Suppose you have 1000 datasets, and for each dataset you need to make a graph and save it as a jpeg. It would take a great deal of time to do this manually, and a mechanism that can repeat the same (or similar) commands any number of times without human intervention would be invaluable. This is where a loop comes in. A plan for the 1000 datasets could be

---

For i is from 1 to 1000:
        Extract dataset i
        Choose appropriate labels for the graph for dataset i
        Make a graph for dataset i
        Save the graph for dataset i
End of loop

Note that this is not R code. It is merely a schematic overview, which is the reason that we put the text in a box and did not use the " > " symbol and the `Courier New` font that we have been using for R code. The sketch involves a loop, meaning that, once the code is syntax correct, R executes 1000 iterations, with the first iteration having $i = 1$, the second iteration $i = 2$, and in the final iteration $i = 1000$. In each iteration, the commands inside the loop are executed.

This plan has only four steps, but, if we want to do more with the data, it may make sense to group certain commands and put them in a function. Suppose we not only want a graph for each dataset, but also to calculate summary statistics and apply a multivariate analysis. We will very quickly end up with 10–15 commands inside the loop, and the code becomes difficult to manage. In such a scenario, using functions can keep the code simple:

For i is from 1 to 1000:
        Extract dataset i
        Execute a function to calculate summary statistics for dataset i.
        Execute a function to make and save a graph for dataset i.
        Execute a function that applies multivariate analysis on dataset i.
End of loop

Each function is a small collection of commands acting on individual datasets. Each function works independently, unaffected by what happens elsewhere, and does only what it has been told to do. There is a mechanism in place to allow only the dataset into the function and to return information for this dataset. Once programmed, the function should work for any dataset. Program it once, and, if all goes according to plan, you never have to think about it again.

Just as a house can be designed to be built in different ways, your plan can take more than one approach. In the sketch above, we created a loop for *i* from 1 to 1000, which, in each iteration, extracts data and passes the data to a function. You can also do it the other way around:

> Execute a function to calculate summary statistics for each dataset.
> Execute a function to make and save a graph for each dataset.
> Execute a function to apply multivariate analysis on each dataset.

Each function will contain a loop in which the data are extracted and subjected to a series of relevant commands. The building of the code depends entirely on personal programming style, length of the code, type of problem, computing time required, and so on.

Before addressing the creation of functions, we focus on loops.

6.2 Loops

If you are familiar with programming languages like FORTRAN, C, C + +, or MATLAB,[3] you are likely to be familiar with loops. Although R has many tools for avoiding loops, there are situations where it is not possible. To illustrate a situation in which a loop saves considerable time, we use a dataset on begging behaviour of nestling barn owls. Roulin and Bersier (2007) looked at nestlings' response to the presence of the mother and the father. Using microphones inside, and a video camera outside, the nests, they sampled 27 nests, studying vocal begging behaviour when the parents bring prey. A full statistical analysis using mixed effects modelling is presented in Roulin and Bersier (2007) and also in Zuur et al. (2009).

For this example, we use "sibling negotiation," defined as the number of calls by the nestlings in the 30-second interval immediately prior to the arrival of a parent, divided by the number of nestlings. Data were collected between 21.30 hours and 05.30 hours on two consecutive nights. The variable ArrivalTime indicates the time at which a parent arrived at the perch with prey.

Suppose that you have been commissioned to write a report on these data and to produce a scatterplot of sibling negotiation versus arrival time for each nest, preferably in jpeg format. There are 27 nests, so you will need to produce,

---

[3] These are just different types of programming languages, similar to R.

and save, 27 graphs. This is not an uncommon type of task. We have been involved in similar undertakings (e.g., producing multiple contour plots for > 75 bird species in the North Sea). Keep in mind that they may ask you to do it all again with a different plotting character or a different title! Note that R has tools to plot 27 scatterplots in a single graph (we show this in Chapter 8), but assume that the customer has explicitly asked for 27 separate jpeg files. This is not something you will not want to do manually.

6.2.1 Be the Architect of Your Code

Before writing the code, you will need to plan and produce an architectural design outlining the steps in your task:

1. Import the data and familiarise yourself with the variable names, using the read.table, names, and str commands.
2. Extract the data of one nest and make a scatterplot of begging negotiation versus arrival time for this subset.
3. Add a figure title and proper labels along the x- and y-axes. The name of the nest should be in the main header.
4. Extract data from a second nest, and determine what modifications to the original graph are needed.
5. Determine how to save the graph to a jpeg file.
6. Write a loop to extract data for nest *i*, plot the data from nest *i*, and save the graph to a jpeg file with an easily recognized name.

If you can implement this algorithm, you are a good architect!

6.2.2 Step 1: Importing the Data

The following code imports the data and shows the variable names and their status. There is nothing new here in terms of R code; the read.table, names, and str commands were discussed in Chapters 2 and 3.

```
> setwd("C:/RBook/")
> Owls <- read.table(file = "Owls.txt", header = TRUE)
> names(Owls)
[1] "Nest"               "FoodTreatment"
[3] "SexParent"          "ArrivalTime"
[5] "SiblingNegotiation" "BroodSize"
[7] "NegPerChick"
> str(Owls)
'data.frame':   599 obs. of  7 variables:
 $ Nest               : Factor w/ 27 levels ...
```

```
$ FoodTreatment      : Factor w/ 2 levels ...
$ SexParent          : Factor w/ 2 levels ...
$ ArrivalTime        : num 22.2 22.4 22.5 22.6 ...
$ SiblingNegotiation: int 4 0 2 2 2 2 18 4 18 0 ...
$ BroodSize          : int 5 5 5 5 5 5 5 5 5 5 ...
$ NegPerChick        : num 0.8 0 0.4 0.4 0.4 0.4 ...
```

The variables Nest, FoodTreatment, and SexParent are defined using alphanumerical values in the ascii file, and therefore R considers them (correctly) as factors (see the output of the str command for these variables).

6.2.3 Steps 2 and 3: Making the Scatterplot and Adding Labels

To extract the data from one nest, you first need to know the names of the nests. This can be done with the unique command

```
> unique(Owls$Nest)
 [1] AutavauxTV        Bochet          Champmartin
 [4] ChEsard           Chevroux        CorcellesFavres
 [7] Etrabloz          Forel           Franex
[10] GDLV              Gletterens      Henniez
[13] Jeuss             LesPlanches     Lucens
[16] Lully             Marnand         Moutet
[19] Murist            Oleyes          Payerne
[22] Rueyes            Seiry           SEvaz
[25] StAubin           Trey            Yvonnand
27 Levels: AutavauxTV Bochet Champmartin ... Yvonnand
```

There are 27 nests, and their names are given above. Extracting the data of one nest follows the code presented in Chapter 3:

```
> Owls.ATV <- Owls[Owls$Nest=="AutavauxTV", ]
```

Note the comma after Owls$Nest = = "AutavauxTV" to select rows of the data frame. We called the extracted data for this nest Owls.ATV, where ATV refers to the nest name. The procedure for making a scatterplot such as that needed to show arrival time versus negotiation behaviour for the data in Owls.ATV was discussed in Chapter 5. The code is as follows.

```
> Owls.ATV <- Owls[Owls$Nest == "AutavauxTV", ]
> plot(x = Owls.ATV$ArrivalTime,
       y = Owls.ATV$NegPerChick,
       xlab = "Arrival Time", main = "AutavauxTV"
       ylab = "Negotiation behaviour)
```

You will be plotting the variable `ArrivalTime` versus `NegPerChick` from the data frame `Owls.ATV`, hence the use of the $ sign. The resulting graph is presented in Fig. 6.1. So far, the procedure requires no new R code.

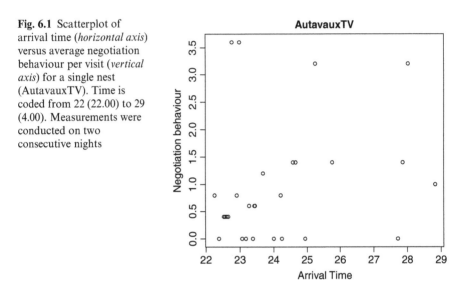

Fig. 6.1 Scatterplot of arrival time (*horizontal axis*) versus average negotiation behaviour per visit (*vertical axis*) for a single nest (AutavauxTV). Time is coded from 22 (22.00) to 29 (4.00). Measurements were conducted on two consecutive nights

6.2.4 Step 4: Designing General Code

To investigate the universality of the code, go through the same procedure for data from another nest. The code for the second nest requires only a small modification; where you entered `AutavauxTV`, you now need `Bochet`.

```
> Owls.Bot <- Owls[Owls$Nest == "Bochet", ]
> plot(x = Owls.Bot$ArrivalTime,
        y = Owls.Bot$NegPerChick,
        xlab = "Arrival Time",
        ylab = "Negotiation behaviour", main = "Bochet")
```

The graph is not shown here. Note that we stored the data from this particular nest in the data frame `Owls.Bot`, where "Bot" indicates "Bochet." If you were to make the same graph for another nest, you need only replace the main title and the name of the data frame and the actual data (the loop will do this for us).

The question is, in as much as you must do this another 25 times, how can you minimise the typing required? First, change the name of the data frame to something more abstract. Instead of `Owls.ATV` or `Owls.Bot`, we used `Owls.i`. The following construction does this.

```
> Owls.i <- Owls[Owls$Nest == "Bochet", ]
> plot(x = Owls.i$ArrivalTime,
      y = Owls.i$NegPerChick, xlab = "Arrival Time",
      ylab = "Negotiation behaviour", main = "Bochet")
```

Instead of a specific name for the extracted data, we used a name that can apply to any dataset and pass it on to the plot function. The resulting graph is not presented here. The name "Bochet" still appears at two places in the code, and they need to be changed each time you work with another dataset. To minimise typing effort (and the chance of mistakes), you can define a variable, Nest.i, containing the name of the nest, and use this for the selection of the data and the main header:

```
> Nest.i <- "Bochet"
> Owls.i <- Owls[Owls$Nest == Nest.i, ]
> plot(x = Owls.i$ArrivalTime, y = Owls.i$NegPerChick,
      xlab = "Arrival Time", main = Nest.i,
      ylab = "Negotiation behaviour")
```

In order to make a plot for another nest, you only need to change the nest name in the first line of code, and everything else will change accordingly.

6.2.5 Step 5: Saving the Graph

You now need to save the graph to a jpeg file (see also the help file of the jpeg function):

1. Choose a file name. This can be anything, for example, ''AnyName.jpg''.
2. Open a jpeg file by typing jpeg(file = ''AnyName.jpg'').
3. Use the plot command to make graphs. Because you typed the jpeg command, R will send all graphs to the jpeg file, and the graphic output will not appear on the screen.
4. Close the jpeg file by typing: dev.off().

You can execute multiple graphing commands in Step 3 (e.g., plot, lines, points, text) and the results of each will go into the jpeg file, until R executes the dev.off (device off) command which closes the file. Any graphing command entered after the dev.off command will not go into the jpeg file, but to the screen again. This process is illustrated in Fig. 6.2.

At this point, you should consider where you want to save the file(s), as it is best to keep them separate from your R working directory. In Chapter 3 we discussed how to set the working directory with the setwd command. We set it to "C:/AllGraphs/" in this example, but you can easily modify this to your own choice.

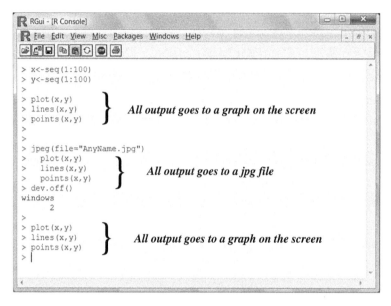

Fig. 6.2 Summary of the `jpeg` and `dev.off` commands. The results of all graphing commands between the `jpeg` and `dev.off` commands are sent to a jpg file. The *x*- and *y*-coordinates were arbitrarily chosen

The final challenge is to create a file name that automatically changes when we change the name of the nest (the variable `Nest.i`). You will need a file name that consists of the nest name (e.g., Bochet) and the file extension jpg. To connect "Bochet" and ".jpg" with no separation between these two strings (i.e., "Bochet.jpg") use the `paste` command:

```
> paste(Nest.i, ".jpg", sep = "")
[1] "Bochet.jpg"
```

The output of the `paste` command is a character string that can be used as the file name. You can store it in a variable and use it in the `jpeg` command. We called the variable `YourFileName` in the code below, and R sends all graphic output created between the `jpeg` and `dev.off` commands to this file.

```
> setwd("C:/AllGraphs/")
> Nest.i <- "Bochet"
> Owls.i <- Owls[Owls$Nest == Nest.i, ]
> YourFileName <- paste(Nest.i, ".jpg", sep="")
> jpeg(file = YourFileName)
> plot(x = Owls.i$ArrivalTime, y = Owls.i$NegPerChick,
     xlab = "Arrival Time", main = Nest.i,
     ylab = "Negotiation behaviour")
> dev.off()
```

Once this code has been executed, you can open the file *Bochet.jpg* in your working directory with any graphic or photo editing package. The help file for the jpeg function contains further information on increasing the size and quality of the jpeg file. Alternative file formats are obtained with the functions bmp, png, tiff, postscript, pdf, and windows. See their help files for details.

6.2.6 Step 6: Constructing the Loop

You still have to modify the variable Nest.i 27 times, and, each time, copy and paste the code into R. Here is where Step 6 comes in, the loop. The syntax of the loop command in R is as follows.

```
for (i in 1 : 27) {
    do something
    do something
    do something
    }
```

"Do something" is not valid R syntax, hence the use of a box. Note that the commands must be between the two curly brackets { and }. We used 27 because there are 27 nests. In each iteration of the loop, the index *i* will take one of the values from 1 to 27. The "do something" represent orders to execute a specific command using the current value of *i*. Thus, you will need to enter into the loop the code for opening the jpeg file, making the plot, and closing the jpeg file for a particular nest. It is only a small extension of the code from Step 5.

On the first line of the code below, we determined the unique names of the nests. On the first line in the loop, we set Nest.i equal to the name of the *i*th nest. So, if *i* is 1, Nest.i is equal to ''AutavauxTV''; *i* = 2 means that Nest.i = ''Bochet''; and, if *i* is 27, Nest.i equals ''Yvonnand'' The rest of the code was discussed in earlier steps. If you run this code, your working directory will contain 27 jpeg files, exactly as planned.

```
> AllNests <- unique(Owls$Nest)
> for (i in 1:27) {
 Nest.i <- AllNests[i]
 Owls.i <- Owls[Owls$Nest == Nest.i, ]
 YourFileName <- paste(Nest.i, ".jpg", sep = "")
 jpeg(file = YourFileName)
 plot(x = Owls.i$ArrivalTime, y = Owls.i$NegPerChick,
      xlab = "Arrival Time",
      ylab = "Negotiation behaviour", main = Nest.i)
 dev.off()
 }
```

Do Exercise 1 in Section 6.6. This is an exercise in creating loops, using a temperature dataset.

6.3 Functions

The principle of a *function* may be new to many readers. If you are not familiar with it, envision a function as a box with multiple holes on one side (for the input) and a single hole on the other side (for the output). The multiple holes can be used to introduce information into the box; the box will act as directed upon the information and feed the results out the single hole. When a function is running properly, we are not really interested in knowing how it obtains the results. We have already used the loess function in Chapter 5. The input consisted of two variables and the output was a list that contained, among other things, the fitted values. Other examples of existing functions are the mean, sd, sapply, and tapply, among others.

The underlying concept of a function is sketched in Fig. 6.3. The input of the function is a set of variables, A, B, and C, which can be vectors, matrices, data frames, or lists. It then carries out the programmed calculations and passes the information to the user.

The best way to learn how to use a function is by seeing some examples.

Fig. 6.3 Illustration of the principle of a function. A function allows for the input of multiple variables, carries out calculations, and passes the results to the user. According to the order in which the variables are entered, A, B, and C are called x, y, and z within the function. This is called positional matching

6.3.1 Zeros and NAs

Before executing a statistical analysis, it is important to locate and deal with any missing values, as they may present some difficulties. Certain techniques, such as linear regression, will remove any case (observation) containing a missing value. Variables with many zeros cause trouble as well, particularly in multivariate analysis. For example, do we say that dolphins and elephants are similar because they are both absent on the moon? For a discussion on double zeros in multivariate analysis, see Legendre and Legendre (1998). In univariate analysis, a response variable with many zeros can also be problematical (See the Zero Inflated Data chapter in Zuur et al., 2009).

We recommend creating a table that gives the number of missing values, and the number of zeros, per variable. A table showing the number of missing values (or zeros) per case is also advisable. The following demonstrates using R code to

create the tables, but before continuing, we suggest that you do Exercise 2 in
Section 6.6, as it guides you through the R code in this section.

Our example uses the vegetation data from Chapter 4. We imported the data
with the read.table command, and used the names command to see the list
of variables:

```
> setwd("C:/RBook/")
> Veg <- read.table(file = "Vegetation2.txt",
                    header = TRUE)
> names(Veg)
 [1] "TransectName" "Samples"     "Transect"
 [4] "Time"         "R"           "ROCK"
 [7] "LITTER"       "ML"          "BARESOIL"
[10] "FallPrec"     "SprPrec"     "SumPrec"
[13] "WinPrec"      "FallTmax"    "SprTmax"
[16] "SumTmax"      "WinTmax"     "FallTmin"
[19] "SprTmin"      "SumTmin"     "WinTmin"
[22] "PCTSAND"      "PCTSILT"     "PCTOrgC"
```

The first four variables contain transect name, transect number, and time of
survey. The column labelled R contains species richness (the number of species)
per observation. The remaining variables are covariates.

Suppose you want a function that takes as input a data frame that contains
the data, and calculates the number of missing values in each variable. The
syntax of such a function is

```
NAPerVariable <- function(X1) {
  D1 <- is.na(X1)
  colSums(D1)
}
```

If you type this code into a text editor and paste it into R, you will see that
nothing happens. The code defines a function with the name NAPerVariable,
but it does not execute the function. This is done with the command

```
> NAPerVariable(Veg[,5:24])
       R     ROCK   LITTER         ML BARESOIL  FallPrec
       0        0        0          0        0         0
 SprPrec  SumPrec  WinPrec   FallTmax  SprTmax   SumTmax
       0        0        0          0        0         0
 WinTmax FallTmin  SprTmin    SumTmin  WinTmin   PCTSAND
       0        0        0          0        0         0
 PCTSILT  PCTOrgC
       0        0
```

We omitted the first four columns of the data frame Veg, as these contain the transect and time information. There appear to be no missing values in the listed variables. Take a closer look at what is going on inside the function. The first, and only, argument of the function is X1. We assume that the variables are in columns and the observations in rows. The command is.na(X1) creates a Boolean matrix of the same dimension as X1, with the value TRUE if the corresponding element of X1 is a missing value and FALSE if not. The colSums function is an existing R function that takes the sum of the elements in each column (variable). Normally, colSums is applied to a data matrix with numbers, but if it is applied to a Boolean matrix, it converts a TRUE to 1, and a FALSE to 0. As a result, the output of colSums(D1) is the number of missing values per variable.

If you replace the colSums command with the rowSums command, the function gives the number of missing values per observation.

6.3.2 Technical Information

There are a few aspects of the function that we need to address: first, the names of the variables used inside the function. Note that we used X1 and D1. You may wonder why the code inside the function runs at all, as X1 seems to come out of the blue. The application here is called *positional matching*. The first and, in this case, only, argument in NAPerVariable, is a subset of the data frame Veg. Inside the function, these data are allocated to X1, because X1 is the first variable in the argument of the function. Hence, X1 contains columns 5 – 24 of the data frame Veg.

The principle of positional matching was illustrated in Fig. 6.1. The external variables A, B, and C are called x, y, and z within the function. R knows that x is A, because both are the first argument in the call to the function. We have already seen this type of action with the arguments in the plot, lines, and loess functions. The reason for changing the variable designations is that you should not use names within a function that also exist outside the function. If you make a programming mistake, for example, if you use D1 <- is.na(X) instead of D1 <- is.na(X1), R will look first inside the function for the values of X. If it does not find this variable inside the function, it will look outside the function. If such a variable exists outside the function, R will happily use it without telling you. Instead of calculating the number of missing values in the variable Veg, it will show you the number of missing values in X, whatever X may be. The convention of using different, or new, names for the variables inside a function applies to all variables, matrices, and data frames used in the function.

A second important aspect of functions is the form in which the resulting information is returned to the user. FORTRAN and C++ users may assume that this is done via the arguments of the function, but this is not the case. It is the information coded for on the final line of the function that is returned. The function `NAPerVariable` has `colSums(D1)` on the last line, so this is the information provided. If you use

```
> H <- NAPerVariable(Veg[ , 4 : 24])
```

`H` will contain the number of missing values in vector format. If the final line of the function is a list, then `H` will be a list as well. In an example presented later in this chapter, we see that this is useful for taking back multiple variables (see also Chapter 3).

As always, you should document your code well. Add comments (with the # symbol) to the function, saying that the data must be in an "observation by variable" format, and that it calculate the number of missing values per column.

You should also ensure that the function will run for every possible dataset that you may enter into it in the future. Our function, for example, will give an error message if the input is a vector (one variable) instead of a matrix; `colSums` only works if the data contain multiple columns (or at least are a matrix). You need to document this, provide an understandable error message, or extend the function so that it will run properly if the input consists of a vector.

6.3.3 A Second Example: Zeros and NAs

The red king crab *Paralithodes camstchaticus* was introduced to the Barents Sea in the 1960 s and 1970 s from its native area in the North Pacific. The leech *Johanssonia arctica* deposits its eggs into the carapace of this crab. The leech is a vector for a trypanosome blood parasite of marine fish, including cod. Hemmingsen et al. (2005) examined a large number of cod for trypanosome infections during annual cruises along the coast of Finnmark in North Norway. We use their data here. The data included the presence or absence of the parasite in fish as well as the number of parasites per fish. Information on the length, weight, age, stage, sex, and location of the host fish was recorded. The familiar `read.table` and `names` functions are used to import the data and show the variable names:

```
> setwd("c:/RBook/")
> Parasite <- read.table(file = "CodParasite.txt",
                         header = TRUE)
> names(Parasite)
[1] "Sample"      "Intensity"   "Prevalence" "Year"
[5] "Depth"       "Weight"      "Length"      "Sex"
[9] "Stage"       "Age"         "Area"
```

Because we already copied and pasted the function `NAPerVariable` into R in Section 6.3.1, there is no need to do this again. To obtain the number of missing values per variable, type

```
> NAPerVariable(Parasite)
Sample  Intensity Prevalence        Year        Depth
     0         57          0           0            0
Weight     Length        Sex       Stage          Age
     6          6          0           0            0
  Area
     0
```

There are 57 missing values in the variable `Intensity`, and 6 in each of the variables `Length` and `Weight`.

In a statistical analysis, we would model the number of parasites as a function of year and length or weight, sex, and location of host fish. This is typically done with generalised linear modelling for count data. Problems may occur if the response variable is zero inflated (too many zeros). Therefore, we need to determine how many zeros are in each variable, especially in `Intensity`. Our first attempt is the function

```
ZerosPerVariable <- function(X1) {
  D1 = (X1 == 0)
  colSums(D1)
}
```

It is similar to the earlier function `NAPerVariable`, except that `D1` is now a matrix with values TRUE if an element of `X1` equals 1, and FALSE otherwise. To execute the function, use

```
> ZerosPerVariable(Parasite)
Sample  Intensity Prevalence        Year        Depth
     0         NA        654           0            0
Weight     Length        Sex       Stage          Age
    NA         NA         82          82           84
  Area
     0
```

There are 654 fish with no parasites, and 82 observations with a value of 0 for Sex. The fact that Sex and Stage have a certain number of observations equal to 0 is a matter of coding; these are nominal variables. So it is not a problem. There are NAs for the variables `Intensity`, `Weight`, and `Length`. This is because the `colSums` function gives NA as output if there is an NA anywhere in the variable. The help file of `colSums` (obtained by typing `? colSums`) shows that the option `na.rm = TRUE` can be added. This leads to:

```
ZerosPerVariable <- function(X1) {
  D1 = (X1 == 0)
```

```
    colSums(D1, na.rm = TRUE)
}
```

Missing values are now ignored because of the na.rm = TRUE option. To execute the new function, we use

```
> ZerosPerVariable(Parasite)
Sample  Intensity Prevalence       Year       Depth
     0        654        654          0           0
Weight     Length        Sex      Stage         Age
     0          0         82         82          84
  Area
     0
```

The output now shows no observations with weight or length equal to 0, and this makes sense. The fact that both Intensity and Prevalence have 654 zeros also makes sense; absence is coded as 0 in the variable Prevalence.

6.3.4 A Function with Multiple Arguments

In the previous section, we created two functions, one to determine the number of missing values per variable and another to find the number of zeros per variable. In this section, we combine them and tell the function to calculate the sum of the number of observations equal to zero or the number of observations equal to NA. The code for the new function is given below.

```
VariableInfo <- function(X1, Choice1) {
    if (Choice1 == "Zeros"){ D1 = (X1 == 0) }
    if (Choice1 == "NAs")   { D1 <- is.na(X1) }
    colSums(D1, na.rm = TRUE)
}
```

The function has two arguments: X1 and Choice1. As before, X1 should contain the data frame, and Choice1 is a variable that should contain either the value "Zeros" or "NAs." To execute the function, use

```
> VariableInfo(Parasite, "Zeros")
Sample  Intensity Prevalence       Year       Depth
     0        654        654          0           0
Weight     Length        Sex      Stage         Age
     0          0         82         82          84
  Area
     0
```

For the missing values, we can use

```
> VariableInfo(Parasite, "NAs")
Sample   Intensity Prevalence      Year       Depth
     0          57          0         0           0
Weight     Length        Sex      Stage         Age
     6           6          0         0           0
  Area
     0
```

As you can see, the output is the same as in the previous section. So, the function performs as we intended. We can also allocate the output of the function to a variable in order to store it.

```
> Results <- VariableInfo(Parasite, "Zeros")
```

If you now type `Results` into the console, you will get the same numbers as above. Figure 6.4 gives a schematic overview of the function up to this point. The function takes as input the data frame `Parasite` and the character string `"Zeros"`, and internally calls them `X1` and `Choice1`, respectively. The function then performs its calculations and the final result is stored in `D1`. Outside the function, the results are available as `Results`. Once everything is perfectly coded and bug free, you can forget about `X1`, `Choice1`, and `D1`, and what is going on inside the function; all that matters is the input and the results.

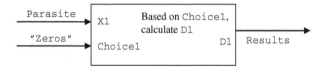

Fig. 6.4 Illustration of the function to calculate the number of zeros or the number of missing values of a dataset. Due to positional matching, the data frame `Parasite` and the argument "Zeros" are called `X1` and `Choice1` within the function

The only problem is that our current function is not robust against user error. Suppose you make a typing mistake, spelling "Zeros" as "zeroos":

```
> VariableInfo(Parasite, "zeroos")
Error in inherits(x, "data.frame"): object "D1" not
found
```

The variable `Choice1` is equal to the nonexistent "zeroos", and therefore none of the commands is executed. Hence, `D1` has no value, and an error message is given on the last line. Another possible mistake is to forget to include a value for the second argument:

```
> VariableInfo(Parasite)
Error in VariableInfo(Parasite): argument "Choice1" is
missing, with no default
```

The variable `Choice1` has no value; the code crashes at the first line. The challenge in making a function is anticipating likely errors. Here, we have seen two silly (but common) mistakes, but the function can be written to provide a safety net for these types of errors.

6.3.5 Foolproof Functions

To make a foolproof function, you have to give it to hundreds of people and ask them all to try it and report any errors, or apply it on hundreds of datasets. Even then, you may be able to crash it. But there are a few common things you can do to make it as stable as possible.

6.3.5.1 Default Values for Variables in Function Arguments

The variable `Choice1` can be given a default value so that if you forget to enter a value for `Choice1`, the function will do the calculations for the default value. This is done with

```
VariableInfo <- function(X1, Choice1 = "Zeros") {
  if (Choice1 == "Zeros") { D1 = (X1 == 0) }
  if (Choice1 == "NAs")   { D1 <- is.na(X1) }
  colSums(D1, na.rm = TRUE)
}
```

The default value is now "Zeros." Executing this function without specifying a value for `Choice1` produces valid output. To test it, type

```
> VariableInfo(Parasite)
```

Sample	Intensity	Prevalence	Year	Depth
0	654	654	0	0
Weight	Length	Sex	Stage	Age
0	0	82	82	84
Area				
0				

To calculate the number of missing values, use as before:

```
> VariableInfo(Parasite, "NAs")
```

In this case, the second `if` command in the function is executed. The output of this command is not shown here. Don't forget to write a help file to document the default value!

6.3.5.2 Misspelling

We also want a function that executes the appropriate code, depending on the value of `Choice1`, and gives a warning message if `Choice1` is not equal to "Zeros" or "NAs". The following code does just that.

```
VariableInfo <- function(X1, Choice1 = "Zeros") {
    if (Choice1 == "Zeros") { D1 = (X1 == 0) }
    if (Choice1 == "NAs")   { D1 <- is.na(X1) }
    if (Choice1 != "Zeros" & Choice1 != "NAs") {
        print("You made a typo") } else {
             colSums(D1, na.rm = TRUE) }
}
```

The third `if` statement will print a message if `Choice1` is not equal to either "Zeros" or "NAs". If one of these conditions is TRUE, then the `colSums` command is executed. To see it in action, type:

```
> VariableInfo(Parasite, "abracadabra")
```

```
[1] "You made a typo"
```

Note that internally the function is doing the following steps.

```
If A then blah blah
If B then blah blah
If C then blah blah, ELSE blah blah
```

A professional programmer will criticise this structure, as each `if` statement is inspected by R, even if the argument is "Zero" and only the first `if` statement is relevant. In this case, this does not matter, as there are only three `if` statements which won't take much time, but suppose there are 1000 `if` statements, only one of which needs to be executed. Inspecting the entire list is a waste of time. The help file for the `if` command, obtained by `? if`, provides some tools to address this situation. In the "See also" section, there is a link to the `ifelse` command. This can be used to replace the first two commands in the function:

```
> ifelse(Choice1 == "Zeros", D1 <- (X1 == 0),
                             D1 <- is.na(X1))
```

If the value of `Choice1` is equal to "Zeros", then the $D1 <-(X1 == 0)$
command is executed, and, in all other situations, it is $D1 <- is.na(X1)$. Not
exactly what we had in mind, but it illustrates the range of options available in
R. In Section 6.4, we demonstrate the use of the `if else` construction to avoid
inspecting a large number of if statements.

Do Exercise 2 in Section 6.6 on creating a new categorical variable
with the `ifelse` command, using the owl data.

6.4 More on Functions and the `if` Statement

In the following we discuss passing multiple arguments out of a function and the
`ifelse` command, with the help of a multivariate dataset. The Dutch govern-
ment institute RIKZ carried out a marine benthic sampling program in the
summer of 2002. Data on approximately 75 marine benthic species were collected at 45 sites on nine beaches along the Dutch coastline. Further information
on these data and results of statistical analyses such as linear regression, generalised additive modelling, and linear mixed effects modelling, can be found in Zuur
et al. (2007, 2009).

The data matrix consists of 45 rows (sites) and 88 columns (75 species and 13
explanatory variables). You could apply multivariate analysis techniques to see
which species co-occur, which sites are similar in species composition, and which
environmental variables are driving the species abundances. However, before
doing any of this, you may want to start simply, by calculating a diversity index
and relating this index to the explanatory variables.

A diversity index means that, for each site, you will characterise the 75 species
with a single value. There are different ways of doing this, and Magurran (2004)
describes various diversity indices. We do not want to engage in a discussion of
which is better. You only need to develop an R function that takes as input an
observation-by-species matrix, potentially with missing values, and a variable
that tells the function which diversity index to calculate. To keep it simple, we
limit the code to three indices. Interested readers can extend this R function and
add their own favourite diversity indices. The three indices we use are:

1. Total abundance per site.
2. Species richness, defined as the number of different species per site.
3. The Shannon index. This takes into account both the presence/absence
   nature of the data and the actual abundance. It is defined by

$$H_i = - \sum_i^m p_{ij} \times \log_{10} p_{ij}$$

p_{ij} is calculated by

$$p_{ij} = \frac{Y_{ij}}{\sum_{j=1}^{n} Y_{ij}}$$

where p_{ij} is the proportion of a particular species j at site i, and m (in the first equation) is the total number of species. The total number of species is n.

6.4.1 Playing the Architect Again

Just as with the previous example presented in this chapter, begin by making a sketch of the tasks to be carried out.

1. Import the data and investigate what you have in terms of types of variables, variable names, dimension of the data, and so on.
2. Calculate total abundance for site 1. Repeat this for site 2. Automate this process, making the code as general as possible. Use elegant and efficient coding.
3. Calculate the different number of species for site 1. Repeat this process for site 2. Automate this process, and make the code as general as possible.
4. Do the same for the Shannon index.
5. Combine the code, and use an if statement to choose between the indices. Use elegant coding.
6. Put all the code in a function and allow the user to specify the data and the diversity index. The function should return the actual index and also indicate which diversity index was chosen (as a string).

In the following, we transform this sketch into fully working R code.

6.4.2 Step 1: Importing and Assessing the Data

Import the RIKZ data, separate the species data from the environmental data, and determine the size of the data with the following R code.

```
> Benthic <- read.table("C:/RBook/RIKZ.txt",
                 header = TRUE)
> Species <- Benthic[ , 2:76]
> n <- dim(Species)
> n
[1] 45 75
```

The first column in the data frame Benthic contains labels, columns 2–76 contain species data, and columns 77–86 are the explanatory variables. The species data are extracted and stored in the data frame `Species`. Its dimension is 45 rows and 75 columns, and these values are obtained and stored in n using the `dim` command. To save space, results of the `names` and `str` command are not shown here; all variables are coded numerically.

6.4.3 Step 2: Total Abundance per Site

Calculate the sum of all species at site 1 by using

```
> sum(Species[1, ], na.rm = TRUE)

[1] 143
```

The total number of species at site 1 is 143. The same can be done for site 2:

```
> sum(Species[2, ], na.rm = TRUE)

[1] 52
```

To avoid typing this command 45 times, construct a loop that calculates the sum of all species per site. Obviously, we need to store these values. The following code does this.

```
> TA <- vector(length = n[1])
> for (i in 1:n[1]){
    TA[i] <- sum(Species[i, ], na.rm = TRUE)
    }
```

The vector `TA` is of length 45 and contains the sum of all species per site:

```
> TA

 [1] 143  52  70 199  67 944 241 192 211 48 35
[12]   1  47  38  10   1  47  73   8  48  6 42
[23]  29   0  43  33  34  67  46   5   7  1  1
[34] 102 352   6  99  27  85   0  19  34 23  0
[45]  11
```

Three sites have no species at all, whereas at one site the total abundance is 944. Note that you must define TA as a vector of length 45 before constructing the loop or TA[i] will give an error message (see the code above). You also need to ensure that the index *i* in the loop is indeed between 1 and 45; T[46] is not defined. Instead of using `length` = 45 in the vector command, we used length = n[1]; remember that the task was to make the code as general as possible. The loop is what we call the brute force approach, as more elegant programming, producing identical results, is given by:

```
> TA <- rowSums(Species, na.rm = TRUE)
> TA
```

```
 [1] 143   52 70 199 67 944 241 192 211 48 35
[12]   1   47 38  10  1  47  73   8  48  6 42
[23]  29    0 43  33 34  67  46   5   7  1  1
[34] 102 352  6  99 27  85   0  19  34 23  0
[45]  11
```

The rowSums command takes the sum for each row. Note that this requires only one line of coding and also involves less computing time (albeit for such a small dataset the difference is very small), and is preferable to the loop.

6.4.4 Step 3: Richness per Site

The number of species at site 1 is given by

```
> sum(Species[1, ] > 0, na.rm = TRUE)
```

```
[1] 11
```

There are 11 different species at site 1. Species[1,] > 0 creates a Boolean vector of length 75 with elements TRUE and FALSE. The function sum converts the value TRUE to 1, and FALSE to 0, and adding these values does the rest.

For site 2, use

```
> sum(Species[2, ] > 0, na.rm = TRUE)
```

```
[1] 10
```

To calculate the richness at each site, create a loop as for total abundance. First define a vector Richness of length 45, then execute a loop from 1 to 45. For each site, richness is determined and stored.

```
> Richness <- vector(length = n[1])
> for (i in 1:n[1]) {
```

```
    Richness[i] <- sum(Species[i, ] > 0, na.rm = TRUE)
    }
> Richness

 [1] 11 10 13 11 10 8   9 8 19 17 6   1 4 3 3
[16]   1  3  3  1   4 3 22 6   0   6 5   4 1 6 4
[31]   2  1  1  3   4 3   5 7   5   0 7 11 3 0 2
```

The elegant approach uses the rowSums command and gives the same result:

```
> Richness <- rowSums(Species > 0, na.rm = TRUE)
> Richness

 [1] 11 10 13 11 10 8   9 8 19 17 6   1 4 3 3
[16]   1  3  3  1   4 3 22 6   0   6 5   4 1 6 4
[31]   2  1  1  3   4 3   5 7   5   0 7 11 3 0 2
```

6.4.5 Step 4: Shannon Index per Site

To calculate the Shannon index, we need only three lines of elegant R code that include the equations of the index:

```
> RS <- rowSums(Species, na.rm = TRUE)
> prop <- Species / RS
> H <- -rowSums(prop * log10(prop), na.rm = TRUE)
> H

 [1]   0.76190639   0.72097224   0.84673524
 [4]   0.53083926   0.74413939   0.12513164
 [7]   0.40192006   0.29160667   1.01888185
[10]   0.99664096   0.59084434   0.00000000
               < Cut to reduce space>
```

We could have used code with a loop instead. The calculation can be done even faster with the function "diversity", which can be found in the vegan package in R. This package is not part of the base installation; to install it, see Chapter 1. Once installed, the following code can be used.

```
> library(vegan)
> H <- diversity(Species)
> H
```

```
        1         2         3         4         5
1.7543543 1.6600999 1.9496799 1.2223026 1.7134443
        6         7         8         9        10
0.2881262 0.9254551 0.6714492 2.3460622 2.2948506
       11        12        13        14        15
1.3604694 0.0000000 0.4511112 0.5939732 0.9433484
       16        17        18        19        20
0.0000000 0.7730166 0.1975696 0.0000000 0.8627246
              < Cut to reduce space>
```

Note that the values are different. The `diversity` help file shows that this function uses the natural logarithmic transformation, whereas we used the logarithm with base 10. The `diversity` help file gives instructions for changing this when appropriate.

A limitation of using the `vegan` package is that this package must be installed on the computer of the user of your code.

6.4.6 Step 5: Combining Code

Enter the code for all three indices and use an `if` statement to select a particular index.

```
> Choice <- "Richness"
> if (Choice == "Richness") {
      Index <- rowSums(Species >0, na.rm = TRUE)}
> if (Choice == "Total Abundance") {
      Index <- rowSums(Species, na.rm = TRUE)  }
> if (Choice == "Shannon")  {
      RS <- rowSums(Species, na.rm = TRUE)
      prop <- Species / RS
      Index <- -rowSums(prop*log10(prop), na.rm = TRUE)}
```

Just change the value of `Choice` to '' `Total Abundance`'' or" `Shannon`'' to calculate the other indices.

6.4.7 Step 6: Putting the Code into a Function

You can now combine all the code into one function and ensure that the appropriate index is calculated and returned to the user. The following code does this.

```
Index.function <- function(Spec, Choice1){
  if (Choice1 == "Richness") {
```

```
    Index <- rowSums(Spec > 0, na.rm = TRUE)}
if (Choice1 == "Total Abundance") {
    Index <- rowSums(Spec, na.rm = TRUE) }
if (Choice1 == "Shannon") {
   RS <- rowSums(Spec, na.rm = TRUE)
   prop <- Spec / RS
   Index <- -rowSums(prop * log10(prop),
                     na.rm = TRUE)}
list(Index = Index, MyChoice = Choice1)
}
```

The `if` statement ensures that only one index is calculated. For small datasets, you could calculate them all, but for larger datasets this is not good practice. Before executing the code, it may be wise to ensure that none of the variables within the function also exists outside the function. If they do, remove them with the `rm` command (see Chapter 1), or quit and restart R. We renamed all input variables so that no duplication of variable names is possible. In order to execute the function, copy the code for the function, paste it into the console, and type the command:

```
> Index.function(Species, "Shannon")
```

```
$Index
```

```
 [1]  0.76190639 0.72097224 0.84673524 0.53083926
 [5]  0.74413939 0.12513164 0.40192006 0.29160667
 [9]  1.01888185 0.99664096 0.59084434 0.00000000
[13]  0.19591509 0.25795928 0.40969100 0.00000000
[17]  0.33571686 0.08580337 0.00000000 0.37467654
[21]  0.37677792 1.23972435 0.62665477 0.00000000
[25]  0.35252466 0.39057516 0.38359186 0.00000000
[29]  0.58227815 0.57855801 0.17811125 0.00000000
[33]  0.00000000 0.12082909 0.08488495 0.43924729
[37]  0.56065567 0.73993117 0.20525195 0.00000000
[41]  0.65737571 0.75199627 0.45767851 0.00000000
[45]  0.25447599
```

```
$MyChoice
```

```
[1] "Shannon"
```

Note that the function returns information from of the final command, which in this case is a `list` command. Recall from Chapter 2 that a `list` allows us to combine data of different dimensions, in this case a variable with 45 values and also the selected index.

Is this function perfect? The answer is no, as can be verified by typing

```
> Index.function(Species, "total abundance")
```

The error message produced by R is

```
Error in Index.function(Species, "total abundance"):
  object "Index" not found
```

Note that we made a typing error in not capitalizing "total abundance". In the previous section, we discussed how to avoid such errors. We extend the function so that it inspects all `if` statements and, if none of them is executed, gives a warning message. We can use the `if else` command for this.

```
Index.function <- function(Spec,Choice1) {
  if (Choice1 == "Richness") {
    Index <- rowSums(Spec > 0, na.rm = TRUE) } else
  if (Choice1 == "Total Abundance") {
    Index <- rowSums(Spec, na.rm = TRUE) } else
  if (Choice1 == "Shannon") {
    RS <- rowSums(Spec, na.rm = TRUE)
    prop <- Spec / RS
    Index <- -rowSums(prop*log(prop),na.rm=TRUE) } else {
      print("Check your choice")
      Index <- NA }
  list(Index = Index, MyChoice = Choice1) }
```

R will look at the first `if` command, and, if the argument is FALSE, it will go to the second `if` statement, and so on. If the variable `Choice1` is not equal to "Richness", "Total Abundance", or "Shannon", the function will execute the command,

```
print("Check your choice")
Index <- NA
```

You can replace the text inside the `print` command with anything appropriate. It is also possible to use the `stop` command to halt R. This is useful if the function is part of a larger calculation process, for example, a bootstrap procedure. See the help files on `stop`, `break`, `geterrmessage`, or `warning`. These will help you to create specific actions to deal with unexpected errors in your code.

Table 6.1 R functions introduced in this chapter

Function	Purpose	Example
jpeg	Opens a jpg file	jpeg(file = "AnyName.jpg")
dev.off	Closes the jpg file	dev.off()
function	Makes a function	z <- function(x, y){ }
paste	Concatenates variables as characters	paste("a", "b", sep = " ")
if	Conditional statement	if (a) { x <-1 }
ifelse	Conditional statement	ifelse (a, x <-1, x <-2)
if elseif	Conditional statement	if (a) { x <-1 } elseif (b) { x <-2 }

6.5 Which R Functions Did We Learn?

Table 6.1 shows the R functions that were introduced in this chapter.

6.6 Exercises

Exercise 1. Using a loop to plot temperature per location.
In Section 6.2, sibling negotiation behaviour was plotted versus arrival time for each nest in the owl data. A graph for each nest was created and saved as a jpg file. Do the same for the temperature data; see Exercise 4.1 for details. The file *temperature.xls* contains temperature observations made at 31 locations (denoted as stations in the spreadsheet) along the Dutch coastline. Plot the temperature data versus time for each station, and save the graph as a jpg file.

Exercise 2. Using the ifelse command for the owl data.
The owl data were sampled on two consecutive nights. If you select the data from one nest, the observations will cover both nights. The two nights differed as to the feeding regime (satiated or deprived). To see observations from a single night, select all observations from a particular nest and food treatment. Use the ifelse and paste functions to make a new categorical variable that defines the observations from a single night at a particular nest. Try rerunning the code from Exercise 1 to make a graph of sibling negotiation versus arrival time for observations of the same nest and night.

Exercise 3. Using the function and if commands with the benthic dataset.
In this exercise we provide the steps for the function that was presented in Section 6.4: the calculation of diversity indices. Read the introductory text in Section 6.4 on diversity indices. Import the benthic data and extract columns 2–76; these are the species.

Calculate total abundance at site 1. Calculate total abundance at site 2. Calculate total abundance at site 3. Calculate the total abundance at site 45. Find a function that can do this in one step (sum per row). Brute force may work as well (`loop`), but is less elegant.

Calculate the total number of *different* species in site 1 (species richness). Calculate species richness for site 2. Do the same for sites 3 and 45. Find a function that can do this in one step.

Create a function using the code for all the diversity indices. Make sure that the user can choose which index is calculated. Ensure that the code can deal with missing values.

If you are brave, add the Shannon index. Apply the same function to the vegetation data.

Chapter 7
Graphing Tools

Chapter 5, the `plot` function was introduced. We demonstrated elementary scatterplots, modifying plotting characters, and adding *x*- and *y*-labels and a main title. In this chapter, we introduce more graphing tools. Not all of them are among our favourites. For example, we have never used pie charts or bar charts. However, these graphs seem to be on the shortlist of so many scientists that we find it necessary to include them in this book. They are discussed in Sections 7.1 and 7.2. Tools to detect outliers—the boxplot and Cleveland dotplot—are presented in Sections 7.3 and 7.4, respectively. We also demonstrate graphs illustrating the mean with lines added to represent the standard error. Scatterplots are further discussed in Section 7.5. Multipanel scatterplots are discussed in Sections 7.6 and 7.7, and advanced tools to display multiple graphs in a single window are presented in Section 7.8.

7.1 The Pie Chart

7.1.1 Pie Chart Showing Avian Influenza Data

We demonstrate the pie chart using the avian influenza dataset from Exercise 1 in Section 3.7. Recall that the data represent the numbers of confirmed human cases of Avian Influenza A/(H5N1) reported to the World Health Organization (WHO). The data for several countries were taken from the WHO website at www.who.int and are reproduced only for educational purposes. We exported the data in the Excel file, *BidFlu.xls*, to a tab-separated ascii file with the name *Birdflucases.txt*. The following code imports the data and presents the usual information.

```
> setwd("C:/RBook/")
> BFCases <- read.table(file = "Birdflucases.txt",
                        header = TRUE)
> names(BFCases)
 [1] "Year"      "Azerbaijan" "Bangladesh"
 [4] "Cambodia"  "China"      "Djibouti"
```

A.F. Zuur et al., *A Beginner's Guide to R*, Use R,
DOI 10.1007/978-0-387-93837-0_7, © Springer Science+Business Media, LLC 2009

```
 [7] "Egypt"      "Indonesia."  "Iraq"
[10] "LaoPDR"     "Myanmar"     "Nigeria"
[13] "Pakistan"   "Thailand"    "Turkey"
[16] "VietNam"


> str(BFCases)


'data.frame': 6 obs. of 16 variables:
$Year      : int 2003 2004 2005 2006 2007 2008
$Azerbaijan: int 0 0 0 8 0 0
$Bangladesh: int 0 0 0 0 0 1
$Cambodia  : int 0 0 4 2 1 0
$China     : int 1 0 8 13 5 3
$Djibouti  : int 0 0 0 1 0 0
$Egypt     : int 0 0 0 18 25 7
$Indonesia.: int 0 0 20 55 42 18
$Iraq      : int 0 0 0 3 0 0
$LaoPDR    : int 0 0 0 0 2 0
$Myanmar   : int 0 0 0 0 1 0
$Nigeria   : int 0 0 0 0 1 0
$Pakistan  : int 0 0 0 0 3 0
$Thailand  : int 0 17 5 3 0 0
$Turkey    : int 0 0 0 12 0 0
$VietNam   : int 3 29 61 0 8 5
```

We have annual data from the years 2003–2008. The first variable contains the years. There are various things we can learn from this dataset. An interesting question is whether the number of bird flu cases has increased over time. We can address this question for individual countries or for the total number of cases. The latter is calculated by

```
> Cases <- rowSums(BFCases[, 2:16])
> names(Cases) <- BFCases[, 1]
> Cases


2003 2004 2005 2006 2007 2008
   4   46   98  115   88   34
```

Columns 2–16 of BFCases contain the information per country. The row-Sums function calculates totals per year and the names function adds the labels 2003–2008 to the variable Cases. (Note that the 34 cases in 2008 is misleading, as this was written halfway through 2008. If this were a proper statistical analysis, the 2008 data would be dropped.) The function for a pie chart in R is pie. It has various options, some of which are illustrated in Fig. 7.1. The pie function requires as input a vector of nonnegative numerical quantities; anything more is optional and deals with labels, colours, and the like.

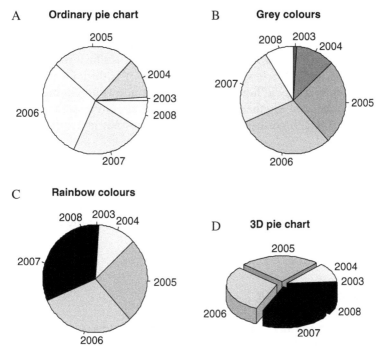

Fig. 7.1 A: Standard pie chart. **B**: Pie chart with clockwise direction of the slices. **C**: Pie chart with rainbow colours (which have been converted to greyscale during the printing process). **D**: Three-dimensional pie chart

Figure 7.1 was made with the following R code.

```
> par(mfrow = c(2, 2), mar = c(3, 3, 2, 1))
> pie(Cases, main = "Ordinary pie chart")                      #A
> pie(Cases, col = gray(seq(0.4, 1.0, length = 6)),
      clockwise = TRUE, main = "Grey colours")                 #B
> pie(Cases, col = rainbow(6), clockwise = TRUE,
      main = "Rainbow colours")                                #C
> library(plotrix)
> pie3D(Cases, labels = names(Cases), explode = 0.1,
      main = "3D pie chart", labelcex = 0.6)                   #D
```

The par function is discussed in the next section. The variable Cases is of length 6 and contains totals per year. The command pie (Cases) creates the pie chart in Fig. 7.1A. Note that the direction of the slices is anticlockwise, which may be awkward, because our variable is time related. We reversed this in the second pie chart (Fig. 7.1B) with the option clockwise = TRUE. We also changed the colours, but, because this book is printed without colour, try this yourself: type in the code and see the colours of the pie charts in panels A–C. Because most of your work is likely to end up in a greyscale paper or report, we recommend using greyscale from the beginning. The only exception is for a PowerPoint presentation, where it is useful to present coloured pie charts. Note that the term "useful" refers to "coloured," rather than to pie charts per se. The main problem with the pie chart is illustrated in Fig. 7.1: Although 2005 and 2006 have the largest slices, it is difficult to determine whether you should stay at home and close the windows and doors to survive the next pandemic, or whether "only" a handful of people were unfortunate enough to contract the disease. The pie chart does not give information on sample size.

Finally, Fig. 7.1D shows a three-dimensional pie chart. Although it now looks more like a real pie, it is, if anything, even less clear in its presentation than the other three graphs. To make this graph, you need to install the package plotrix. The function pie3D has many options, and we suggest that you consult its help file to improve the readability of labels.

7.1.2 The par Function

The par function has an extensive list of graph parameters (see ? par) that can be changed. Some options are helpful; others you may never use.

The mfrow =c (2, 2) creates a graphic window with four panels. Changing the c (2, 2) to c (1, 4) or c (4, 1) produces a row (or column) of four pie charts. If you have more than four graphs, for instance 12, use mfrow =c (3, 4), although now things can become crowded.

The mar option specifies the amount of white space around each graph (each pie chart in this case). The white space is defined by the number of lines of margin at the four sides; bottom, left, top, and right. The default values are, respectively, c (5, 4, 4, 2) + 0.1. Increasing the values gives more white space. Using trial and error, we chose c (3, 3, 2, 1).

A problem arises with the par function if you execute the code for the four pie charts above and, subsequently, make another graph. R is still in the 2 × 2 mode, and will overwrite Figure 7.1A, leaving the other three graphs as they are. The next graph will overwrite panel B, and so on. There are two ways to avoid this. The first option is simply to close the four-panel graph in R before making a new one. This is a single mouse click. The alternative is a bit more programming intensive:

```
> op <- par(mfrow = c(2, 2), mar = c(3, 3, 2, 1))
> pie(Cases, main = "Ordinary pie chart")
> pie(Cases, col = gray(seq(0.4, 1.0, length = 6)),
        clockwise = TRUE, main = "Grey colours")
> pie(Cases, col = rainbow(6), clockwise = TRUE,
        main = "Rainbow colours")
```

```
> pie3D(Cases, labels = names(Cases), explode = 0.1,
      main = "3D pie chart", labelcex = 0.6)
> par(op)
```

The graph parameter settings are stored in the variable op on the first line. The graphs are made as before, and the last line of code converts to the default settings. Any new graph created after the par(op) command will be plotted as if the par function had not been used. This is useful if you need to create many graphs in sequence. It is neat programming, but takes more typing. It is often tempting to be lazy and go for the first approach. However, for good programming practice, we recommend making the extra effort. You will also see this style of programming in the help files.

Do Exercise 1 in Section 7.10 using the pie function.

7.2 The Bar Chart and Strip Chart

We give two examples of the bar chart, another type of graph that is not part of our toolbox. In the first example, we continue with the avian influenza data and present a bar chart showing the total number of bird flu cases and deaths per year. In the second example, a marine benthic dataset is used, with mean values per beach plotted as bars. In the last section, we show a strip chart to visualise similar information.

7.2.1 The Bar Chart Using the Avian Influenza Data

In the previous section, an avian influenza dataset was used to create pie charts showing the total number of cases per year. In addition to bird flu cases, the number of deaths is also available and can be found in the tab-separated ascii file, *Birdfludeaths.txt*. The data are loaded with the commands:

```
> BFDeaths <- read.table(file = "Birdfludeaths.txt",
                             header = TRUE)
> Deaths <- rowSums(BFDeaths[, 2:16])
> names(Deaths) <- BFDeaths[, 1]
> Deaths

2003 2004 2005 2006 2007 2008
   4   32   43   79   59   26
```

The data are structured in the same manner as the bird flu cases. We can visualise the change in the number of cases over time, and then compare number of cases to deaths.

The bar chart in Fig. 7.2A shows the change in the number of cases over time using the data from the variable `Cases` (see Section 7.1 for code to calculate `Cases`). Recall that `Cases` has six values with the labels 2003–2008. Each year is presented as a vertical bar. This graph is more useful than the pie chart, as we can read the absolute values from the *y*-axis. However, a great deal of ink and space is consumed by only six values.

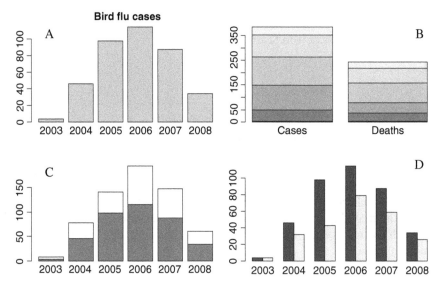

Fig. 7.2 A: Standard bar chart showing the annual number of bird flu cases. **B**: Stacked bar chart showing the accumulated totals per year for cases and deaths (note that values for 2003 can hardly be seen). **C**: Stacked cases (*grey*) and deaths (*white*) per year. **D**: Number of cases and deaths per year represented by adjoining bars

The first two lines of the code below were used to make the bar chart in panel A. The remaining code is for panels B–D:

```
> par(mfrow = c(2, 2), mar = c(3, 3, 2, 1))
> barplot(Cases , main = "Bird flu cases")              #A
> Counts <- cbind(Cases, Deaths)
> barplot(Counts)                                        #B
> barplot(t(Counts), col = gray(c(0.5, 1)))              #C
> barplot(t(Counts), beside = TRUE)                      #D
```

In panels **B–D**, we used the combined data for cases and deaths; these are called Counts and are of dimension 6 × 2:

```
> Counts
        Cases   Deaths
2003      4        4
2004     46       32
2005     98       43
2006    115       79
2007     88       59
2008     34       26
```

In panel B the bars represent data for each year. The graph gives little usable information. Also, years with small numbers (e.g., 2003) are barely visible. To produce panel C, we took the transposed values of Counts using the function t, making the input for the barplot function a matrix of dimension 2 × 6.

```
> t(Counts)
        2003 2004 2005 2006 2007 2008
Cases      4   46   98  115   88   34
Deaths     4   32   43   79   59   26
```

Although you see many such graphs in the literature, they can be misleading. If you compare the white boxes with one another, your eyes tend to compare the values along the *y*-axis, but these are affected by the length of the grey boxes. If your aim is to show that in each year there are more cases than deaths, this graph may be sufficient (comparing compositions). Among the bar charts, panel D is probably the best. It compares cases and deaths within each year, and, because there are only two classes per year, it is also possible to compare cases and deaths among years.

7.2.2 A Bar Chart Showing Mean Values with Standard Deviations

In Chapter 27 of Zuur et al. (2007), core samples were taken at 45 stations on nine beaches along the Dutch coastline. The marine benthic species were determined in each sample with over 75 identified. In Chapter 6, we developed a function to calculate species richness, the number of different species. The file *RIKZ2.txt* contains the richness values for the 45 stations and also a column identifying the beach.

The following R code imports the data and calculates the mean richness and standard deviation per beach. The tapply function was discussed in Chapter 4[1].

---

[1] Note that we could have omitted the text INDEX = and FUN = .

```
> setwd("C:/RBook/")
> Benthic <- read.table(file = "RIKZ2.txt",
                          header = TRUE)
> Bent.M <- tapply(Benthic$Richness,
                INDEX = Benthic$Beach, FUN = mean)
> Bent.sd <- tapply(Benthic$Richness,
                INDEX = Benthic$Beach, FUN = sd)
> MSD <- cbind(Bent.M, Bent.sd)
```

The variable Bent.M contains the mean richness values, and Bent.sd the standard deviation, for each of the nine beaches. We combined them in a matrix MSD with the cbind command. The values are as follows:

```
> MSD

  Bent.M Bent.sd
1   11.0 1.224745
2   12.2 5.357238
3    3.4 1.816590
4    2.4 1.341641
5    7.4 8.532292
6    4.0 1.870829
7    2.2 1.303840
8    4.0 2.645751
9    4.6 4.393177
```

To make a graph in which the mean values are plotted as a bar and the standard deviations as vertical lines extending above the bars (Fig. 7.3A) use the following procedure. For the graph showing mean values, enter

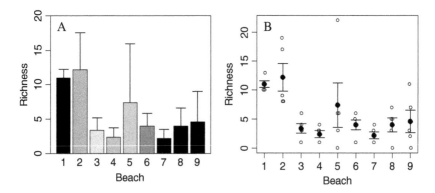

Fig. 7.3 A: Bar chart showing the benthic data. Mean values are represented by the *bars* with a *vertical line* showing standard deviations. The colours were changed to greyscale during the printing process. **B**: Strip chart for the raw data. The mean value per beach is plotted as a *filled dot*, and the *lines* represent the mean $+/-$ the standard error

```
> barplot(Bent.M)
```

Add labels and perhaps some colour for interest:

```
> barplot(Bent.M, xlab = "Beach", ylim = c(0, 20),
          ylab = "Richness", col = rainbow(9))
```

The vertical lines indicating standard deviations are added using the function arrows to draw an arrow between two points with coordinates (x_1, y_1) and (x_2, y_2). Telling R to draw an arrow between the points (x, y_1) and (x, y_2), will produce a vertical arrow, as both points have the same x-value. The y_1-value is the mean, and the y_2-value is the mean plus the standard deviation. The x is the coordinate of the midpoint of a bar. The following code obtains these values and creates Fig. 7.3A.

```
> bp <- barplot(Bent.M, xlab = "Beach", ylim = c(0,20),
                ylab = "Richness", col = rainbow(9))
> arrows(bp, Bent.M, bp, Bent.M + Bent.sd, lwd = 1.5,
         angle = 90, length = 0.1)
> box()
```

It is the bp <-barplot(Bent.M, ...) that helps us out. The best way to understand what it does is by typing:

```
> bp
```

```
        [,1]
 [1,]  0.7
 [2,]  1.9
 [3,]  3.1
 [4,]  4.3
 [5,]  5.5
 [6,]  6.7
 [7,]  7.9
 [8,]  9.1
 [9,] 10.3
```

They are the midpoints along the x-axis of each bar, which are used as input for the arrows function. The angle = 90 and length = 0.1 options change the head of the arrow into a perpendicular line. The lwd stands for line width with 1 as the default value. The box function draws a box around the graph. Run the code without it and see what happens.

7.2.3 The Strip Chart for the Benthic Data

In the previous section, a marine benthic dataset was used, and the mean species richness values per beach were presented as bars with a line representing standard deviation. Section 7.4 in Zar (1999) contains a discussion of when to present the standard deviation, standard error, or twice the standard error (assuming a large sample). It is relatively easy to produce a graph with the raw data, mean values, and either the standard deviation or standard error around the mean. An example is given in Fig. 7.3B. Instead of using the plot function, we used the stripchart function. The open dots show the raw data. We have added random jittering (variation) to distinguish observations with the same value, which would otherwise coincide. The filled dots are the mean values per beach, and were calculated in the previous section. We illustrate the standard errors, which are calculated by dividing the standard deviation by the square root of the sample size (we have five observations per beach). In R, this is done as follows.

```
> Benth.le <- tapply(Benthic$Richness,
              INDEX = Benthic$Beach, FUN = length)
> Bent.se <- Bent.sd / sqrt(Benth.le)
```

The variable Bent.se now contains the standard errors. Adding the lines for standard error to the graph is now a matter of using the arrow function; an arrow is drawn from the mean to the mean plus the standard error, and also from the mean to the mean minus the standard error. The code is below.

```
> stripchart(Benthic$Richness ~ Benthic$Beach,
        vert = TRUE, pch = 1, method = "jitter",
        jit = 0.05, xlab = "Beach", ylab = "Richness")
> points(1:9, Bent.M, pch = 16, cex = 1.5)
> arrows(1:9, Bent.M,
          1:9, Bent.M + Bent.se, lwd = 1.5,
          angle = 90, length = 0.1)
> arrows(1:9, Bent.M,
          1:9, Bent.M - Bent.se, lwd = 1.5,
          angle = 90, length = 0.1)
```

The options in the stripchart function are self-explanatory. Change them to see what happens. The points function adds the dots for the mean values. Instead of the stripchart function, you can also use the plot function, but it does not have a method = "jitter" option. Instead you can use jitter (Benthic$Richness). Similar R code is given in Section 6.1.3 in Dalgaard (2002).

Do Exercise 2 in Section 7.10. This is an exercise in the barchart and stripchart functions using vegetation data.

7.3 Boxplot

7.3.1 Boxplots Showing the Owl Data

The boxplot should most often be your tool of choice, especially when working with a continuous numerical response (dependent) variable and categorical explanatory (independent) variables. Its purpose is threefold: detection of outliers, and displaying heterogeneity of distribution and effects of explanatory variables. Proper use of this graphing tool, along with the Cleveland dotplot (which is described fully in Section 7.4), can provide a head start on analysis of data.

In Chapter 6 we used a dataset on owl research. Roulin and Bersier (2007) looked at how nestlings respond to the presence of the father and of the mother. Using microphones inside and a video outside the nests, they sampled 27 nests, and studied vocal begging behaviour when the parents brought prey. Sampling took place between 21.30 hours and 05.30 hours on two nights. Half the nests were food deprived and the other half food satiated (this was reversed on the second night). The variable ArrivalTime shows the time when a parent arrived at the perch with prey. "Nestling negotiation" indicates the average number of calls per nest.

One of the main questions posed is whether there is a feeding protocol effect and a sex of parent effect. The analysis requires mixed effects modelling techniques and is fully described in Zuur et al. (2009). Before doing any complicated statistics, it is helpful to create boxplots. A boxplot for the nestling negotiation data is easily made using the boxplot function seen in Chapter 1. In Chapter 6, we showed the output of the names and str functions for the owl data, and do not repeat it here.

```
> setwd("C:/RBook/")
> Owls <- read.table(file = "Owls.txt", header = TRUE)
> boxplot(Owls$NegPerChick)
```

The resulting graph is presented in Fig. 7.4. A short description of the boxplot construction is given in the figure labelling. There are five potential outliers, indicating that further investigation is required.

Figure 7.5 illustrates possible effects of sex of the parent (panel A), food treatment (panel B), and the interaction between sex of the parent and food treatment (panels C and D). Because the variable names are long, they are not completely displayed in panel C. We reproduced the boxplots from panel C in panel D and added labels using the names option. Results indicate that there is a possible food treatment effect. The interaction is not clear, which was confirmed by formal statistical analysis. The R code to make Fig. 7.4 is given below. Panels C and D were produced with the SexParent * FoodTreatment construction. The code is self-explanatory.

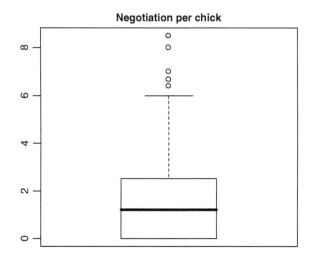

Fig. 7.4 Boxplot of owl nestling negotiation. The *thick horizontal line* is the median; the box is defined by the 25th and 75th percentiles (*lower* and *upper* quartile). The difference between the two is called the spread. The *dotted line* has a length of 1.5 times the spread. (The length of the line pointing up is shorter if the values of the points are smaller than the 75th percentile + 1.5 × spread, and similar for the line pointing downwards. This explains why there is no line at the bottom of the box.) All points outside this range are potential outliers. See Chapter 4 in Zuur et al. (2007) for a discussion of determining if such points are indeed outliers. Note that in this case the 25th percentile is also the smallest value (there are many zero values)

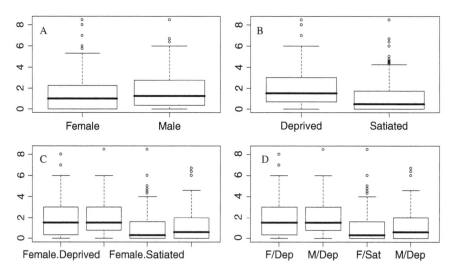

Fig. 7.5 A: Boxplot of owl nestling negotiation conditional on sex of the parent. **B**: Boxplot of owl nestling negotiation conditional on food treatment. **C**: Boxplot of owl nestling negotiation conditional on sex of the parent and food treatment. **D**: Same as panel C, with added labels

```
> par(mfrow = c(2,2), mar = c(3, 3, 2, 1))
> boxplot(NegPerChick ~ SexParent, data = Owls)
> boxplot(NegPerChick ~ FoodTreatment, data = Owls)
> boxplot(NegPerChick ~ SexParent * FoodTreatment,
          data = Owls)
> boxplot(NegPerChick ~ SexParent * FoodTreatment,
        names = c("F/Dep", "M/Dep", "F/Sat", "M/Sat"),
        data = Owls)
```

Sometimes getting all the labels onto a boxplot calls for more creativity. For example, Fig. 7.6 shows a boxplot of nestling negotiation conditional on nest. There are 27 nests, all with long names. If we had entered

```
> boxplot(NegPerChick ~ Nest, data = Owls)
```

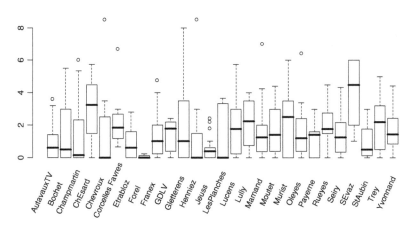

Fig. 7.6 Boxplot of owl nestling negotiation conditional on the 27 nests. The shape of the boxplot suggests that there may be a nest effect, suggesting further analysis by mixed effects models

only a few of the labels would be shown. The solution was to create the boxplot without a horizontal axis line and to put the labels in a small font, at an angle, under the appropriate boxplot. This sounds complicated, but requires only three lines of R code.

```
> par(mar = c(2, 2, 3, 3))
> boxplot(NegPerChick ~ Nest, data = Owls,
          axes = FALSE, ylim = c (-3, 8.5))
> axis(2, at = c(0, 2, 4, 6, 8))
> text(x = 1:27, y = -2, labels = levels(Owls$Nest),
        cex = 0.75, srt = 65)
```

Because we used the option axes = FALSE, the boxplot function drew the boxplot without axes lines. The ylim specifies the lower and upper limits of the vertical axis. Instead of using limits from 0 to 8.5, we used –3 to 8.5. This allowed us to put the labels in the lower part of the graph (Fig. 7.6).

The axis function draws an axis. Because we entered 2 as the first argument, the vertical axis on the left is drawn, and the at argument specifies where the tick marks should be. The text command places all the labels at the appropriate coordinates. The cex argument specifies the font size (1 is default) and srt defines the angle. You will need to experiment with these values and choose the most appropriate settings.

7.3.2 Boxplots Showing the Benthic Data

Recall that in the marine benthic dataset, species richness was measured at nine beaches. We now make a boxplot for each beach (Fig. 7.7). Note that there are only five observations per beach. Because this is a low number for boxplots, we want to add information on sample size per beach to the graph. One option is to specify the varwidth = TRUE option in the boxplot function to make the width of each box proportional to the number of observations on the beach. However, we instead choose to add the number of samples per beach inside each box. First, we need to obtain the sample size per beach using the following R code.

```
> setwd("C:/RBook/")
> Benthic <- read.table(file = "RIKZ2.txt",
                        header= TRUE)
> Bentic.n <- tapply(Benthic$Richness, Benthic$Beach,
                     FUN = length)
> Bentic.n

1 2 3 4 5 6 7 8 9
5 5 5 5 5 5 5 5 5
```

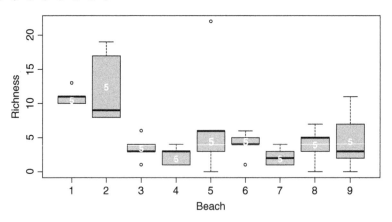

Fig. 7.7 Conditional boxplot using species richness as the dependent variable and beach as the conditioning variable. Number of observations per beach is shown inside each box

The `tapply` function calculates the number of observations per beach, 5, and stores them in the variable `Benthic.n`. The boxplot is created with the command

```
> boxplot(Richness ~ Beach, data = Benthic,
        col = "grey", xlab = "Beach", ylab = "Richness")
```

There is no new code here. The problem is placing the numbers of the variable `Benthic.n` inside the boxplot, preferably in the centre (which is not necessarily the median). Recall that the box is specified by the upper and lower quartiles. Adding half the value of the spread (upper hinge minus lower hinge) to the value of the lower hinge will put us vertically centred in the boxplot. Fortunately, all these values are calculated by the `boxplot` function and can be stored in a list by using

```
> BP.info <- boxplot(Richness ~ Beach, data = Benthic,
                col = "grey", xlab = "Beach",
                ylab = "Richness")
```

The list `BP.info` contains several variables, among them `BP.info $stats`. The `boxplot` help file will tell you that the second row of `$stats` contains the values of the lower hinges (for all beaches), and the fourth row shows the upper hinges. Hence, the midpoints (along the vertical axes) for all beaches are given by:

```
> BP.midp <- BP.info$stats[2, ] +
        (BP.info$stats[4, ] - BP.info$stats[2,]) / 2
```

It is now easy to place the numbers in `Bentic.n` inside the boxplot:

```
> text(1:9, BP.midp, Bentic.n, col = "white", font = 2)
```

We can put any text into the boxplot with this construction. For longer strings, you may want to rotate the text 90 degrees.

The `boxplot` function is very flexible and has a large number of attributes that can be changed. Have a look at the examples in the help files of `boxplot` and `bxp`.

 Do Exercises 3 and 4 in Section 7.10. These are exercises in the `boxplot` function using the vegetation data and a parasite dataset.

7.4 Cleveland Dotplots

Dotplots, also known as Cleveland dotplots, are excellent tools for outlier detection. See Cleveland (1993), Jacoby (2006), or Zuur et al. (2007, 2009) for examples.

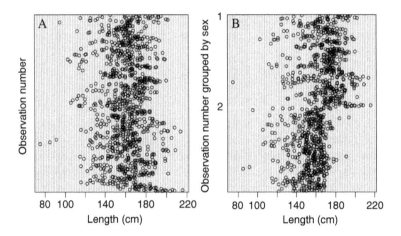

Fig. 7.8 A: Cleveland dotplot showing deer length. The *x*-axis shows the length value and the *y*-axis is the observation number (imported from the ascii file). The first observation is at the bottom of the *y*-axis. **B**: As panel **A**, but with observations grouped according to sex. There may be correlation between length and sex.

Figure 7.8 contains two dotplots for the deer dataset (Vicente et al., 2006), which was used in Section 4.4. Recall that the data were from multiple farms, months, years, and sexes. One of the aims of the study was to assess the relationship between the number of *E. cervi* parasites in deer and the length of the animal. Before doing any analysis, we should inspect each continuous variable in the dataset for outliers. This can be done with a boxplot or with a Cleveland dotplot. Figure 7.8A shows a Cleveland dotplot for the length of the animals. The majority of the animals are around 150 centimetres in length, but there are three animals that are considerably smaller (around 80 centimetres). As a consequence, applying a generalised additive model using length as a smoother may result in larger confidence bands at the lower end of the length gradient.

You can extend a Cleveland dotplot by grouping the observations based on a categorical variable. This was done in Fig. 7.8B; the length values are now grouped by sex. Note that one sex class is clearly larger. The goal of the study was to model the number of parasites (*E. cervi*) as a function of length, sex, year, and farm, in order to determine which of the explanatory (independent) variables is the crucial factor. However, it is difficult to say which explanatory variable is important if there are correlations among the variables. Such a situation is called collinearity. In this case, visualizing length versus sex is useful and can be done with a boxplot in which length is plotted conditional on sex, or with the Cleveland dotplot (Fig. 7.8B).

The graphs were created using the R function, `dotchart`. Function `dotchart2` in the package `Hmisc` (which is not part of the base installation) can produce more sophisticated presentations. We limit our discussion to `dotchart`. The data are imported with the following two lines of code.

```
> setwd("C:/RBook/")
> Deer <- read.table("Deer.txt", header = TRUE)
```

We have seen the output of the names and str commands in Section 4.4, and this information is not repeated. The Cleveland dotplot in Fig. 7.8A is produced with the following R code.

```
> dotchart(Deer$LCT, xlab = "Length (cm)",
           ylab = "Observation number")
```

The dotchart function has various options. The groups option allows grouping the data by categorical variable:

```
> dotchart(Deer$LCT, groups = factor(Deer$Sex))
```

```
Error in plot.window(xlim, ylim, log, asp, ...) :
      need finite 'ylim' values
```

The variable Sex has missing values (type Deer $Sex in the R console to view them), and, as a result, the dotchart function stops and produces an error message. The missing values can easily be removed with the following code.

```
> Isna <- is.na(Deer$Sex)
> dotchart(Deer$LCT[!Isna],
           groups = factor(Deer$Sex[!Isna]),
         xlab = "Length (cm)",
         ylab = "Observation number grouped by sex")
```

The is.na function produces a vector of the same length as Sex, with the values TRUE and FALSE. The ! symbol reverses them, and only the values for which Sex is not a missing value are plotted. Note that we used similar code in Chapter 3. If you want to have the two Cleveland dotplots in one graph, put the par (mfrow = c (1, 2)) in front of the first dotchart.

7.4.1 Adding the Mean to a Cleveland Dotplot

Cleveland dotplots are a good alternative to boxplots when working with small sample sizes. Figure 7.9A shows a Cleveland dotplot of the benthic data used earlier in this chapter. Recall that there are five observations per beach. The right graph shows the same information with the mean value for each beach added. This graph clearly shows at least one "suspicious" observation. The code is basic; see below. The first three commands import the data, with Beach defined as a factor. A graph window with two panels is created with the par function. The first dotchart command follows that of the deer data. To the second dotchart command, we have added the gdata and gpch options.

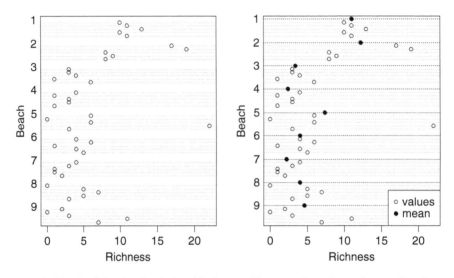

Fig. 7.9 Cleveland dotplots for the benthic data. **A**: The *vertical axis* shows the sampling sites, grouped by beach, and the *horizontal axis* the richness values. **B**: Same as A, with mean values per beach added

The g stands for group, and the `gdata` attribute is used to overlay a summary statistic such as the median, or, as we do here with the `tapply` function, the mean. Finally, the `legend` function is used to add a legend. We discuss the use of the `legend` function in more detail later in this chapter.

```
> setwd("C:/RBook/")
> Benthic <- read.table(file = "RIKZ2.txt",
                        header = TRUE)
> Benthic$fBeach <- factor(Benthic$Beach)
> par(mfrow = c(1, 2))
> dotchart(Benthic$Richness, groups = Benthic$fBeach,
          xlab = "Richness", ylab = "Beach")
> Bent.M<-tapply(Benthic$Richness, Benthic$Beach,
          FUN = mean)
> dotchart(Benthic$Richness, groups = Benthic$fBeach,
          gdata = Bent.M, gpch = 19, xlab = "Richness",
          ylab = "Beach")
> legend("bottomright", c("values", "mean"),
          pch = c(1, 19), bg = "white")
```

 Do Exercises 5 and 6 in Section 7.10 creating Cleveland dotplots for the owl data and for the parasite data.

7.5 Revisiting the plot Function

7.5.1 The Generic plot Function

The most frequently used plotting command is plot, which was introduced in
Chapter 5. It is an intuitive function, recognising what you intend to plot. R is
an object-oriented language: the plot function looks at the object with which it
is presented, establishes the object's class, and recruits the appropriate plotting
method for that object. To see the methods available for a function (i.e., plot),
enter

```
> methods(plot)
```

```
 [1] plot.acf*            plot.data.frame*    plot.Date*
 [4] plot.decomposed.ts*  plot.default        plot.dendrogram*
 [7] plot.density         plot.ecdf           plot.factor*
[10] plot.formula*        plot.hclust*        plot.histogram*
[13] plot.HoltWinters*    plot.isoreg*        plot.lm
[16] plot.medpolish*      plot.mlm            plot.POSIXct*
[19] plot.POSIXlt*        plot.ppr*           plot.prcomp*
[22] plot.princomp*       plot.profile.nls*   plot.spec
[25] plot.spec.coherency  plot.spec.phase     plot.stepfun
[28] plot.stl*            plot.table*         plot.ts
[31] plot.tskernel*       plot.TukeyHSD
   Non-visible functions are asterisked
```

These are the existing plotting functions, and are only those available in
the default packages. All these functions can be called with the plot
function. For example, if you do a principal component analysis (PCA)
and want to print the results, it is not necessary to use the plot.prin-
comp, as the plot function will recognise that you conducted a PCA, and
will call the appropriate plotting function. Another example is the following code.

```
> setwd("C:/RBook/")
> Benthic <- read.table(file = "RIKZ2.txt",
                        header = TRUE)
> Benthic$fBeach <- factor(Benthic$Beach)
> plot(Benthic$Richness ~ Benthic$fBeach)
```

The first three lines import the benthic dataset used earlier in this chapter
and define the variable Beach as a factor. The plot function sees the
formula Benthic$Richness ~ Benthic$fBeach, and produces a boxplot rather than a scatterplot (see the help file of plot.factor). If the
argument in the plot function is a data frame, it will produce a pair plot (see
Section 7.6).

7.5.2 *More Options for the plot Function*

In Chapter 5, we discussed the use of the `plot` function to plot two continuous variables against each other and also showed how to change the characters and colours. But there are many additional options, some of which we present in the remaining part of this section. We use the benthic data to demonstrate once again producing a scatterplot of two continuous variables (Fig. 7.10A). The graph was obtained with the following code.

```
> plot(y = Benthic$Richness, x = Benthic$NAP,
     xlab = "Mean high tide (m)",
     ylab = "Species richness", main = "Benthic data")
> M0 <- lm(Richness ~ NAP, data = Benthic)
> abline(M0)
```

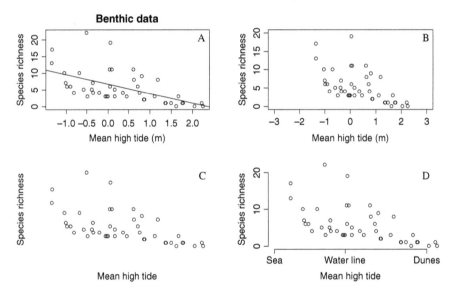

Fig. 7.10 A: Scatterplot of species richness versus NAP (mean high tide levels) with a linear regression line added. **B**: Same as panel **A**, with the *x*- and *y*-ranges set using the `xlim` and `ylim` functions. **C**: Same as panel **A**, but without axes lines. **D**: Same as panel **A**, with modified tick marks along the *y*-axis and character strings along the *x*-axis. Note that the sites are from an intertidal area, hence the negative values of mean high tide

The new addition is the `lm` and `abline` functions. Without going into statistical detail, the `lm` applies linear regression in which species richness is modelled as a function of NAP, the results are stored in the list M0, and the `abline` function superimposes the fitted line. Note that this only works if there is a single explanatory variable (otherwise, plotting the results in a two-dimensional graph becomes difficult), and if the `abline` function is executed following the `plot` function.

The plot function can easily be extended to add more detail to the graph by giving it extra arguments. Some of the most frequently used arguments are given in the table below.

Argument	What does it do?
main	Adds a title to the graph
xlab, ylab	Labels the x- and y- axis
xlim, ylim	Sets limits to the axes
log	Log = "x", log = "y", log = "xy" creates logarithmic axes
type	Type = "p", "l", "b", "o", "h", "s", "n" for plotting points, lines, points connected by lines, points overlaid by lines, vertical lines from points to the zero axis, steps, or only the axes

We have previously illustrated the xlab and ylab attributes. The xlim and ylim specify the ranges along the *x*- and *y*-axes. Suppose that you wish to set the range of the horizontal axis from –3 to 3 metres and the range along the vertical axis from 0 to 20 species. Use

```
> plot(y = Benthic$Richness, x = Benthic$NAP,
    xlab = "Mean high tide (m)",
    ylab = "Species richness",
    xlim = c(-3, 3), ylim = c(0,20))
```

The xlim argument has to be of the form c(x_1, x_2), with numerical values for x_1 and x_2. The same holds for the ylim argument. The results are shown in Fig. 7.10B.

Panels C and D in Fig. 7.10 show other options. Panel C does not contain axes lines. The R code is as follows.

```
> plot(y = Benthic$Richness, x = Benthic$NAP,
    type = "n", axes = FALSE,
    xlab = "Mean high tide",
    ylab = "Species richness")
> points(y = Benthic$Richness, x = Benthic$NAP)
```

The type = n produces a graph without points, and, because we use axes = FALSE, no axes lines are plotted. We begin with a blank window with only the labels. The points function superimposes the points onto the graph (note that you must execute the plot function prior to the points function or an error message will result).

In panel C, we basically told R to prepare a graph window, but not to plot anything. We can then proceed, in steps, to build the graph shown in Panel D. The axis function is the starting point in this process. It allows specifying the position, direction, and size of the tick marks as well as the text labelling them.

```
> plot(y = Benthic$Richness, x = Benthic$NAP,
    type = "n", axes = FALSE, xlab = "Mean high tide",
    ylab = "Species richness",
    xlim = c(-1.75,2), ylim = c(0,20))
> points(y = Benthic$Richness, x = Benthic$NAP)
> axis(2, at = c(0, 10, 20), tcl = 1)
> axis(1, at = c(-1.75, 0,2),
        labels = c("Sea", "Water line", "Dunes"))
```

The first two lines of code are identical to those for panel C. The `axis (2, .. .)` command draws the vertical axis line and inserts tick marks of length 1 (the default value is –0.5) at the values 0, 10, and 20. Setting `tcl` to 0 eliminates tick marks. Tick marks pointing outwards are obtained by a negative `tcl` value; a positive value gives inward pointing tick marks. The `axis (1, ...)` command draws the horizontal axis, and, at the values –1.75, 0, and 2, adds the character strings Sea, Water line, and Dunes. See the `axis` help file for further graphing facilities.

 Do Exercise 7 in Section 7.10. This is an exercise in the `plot` and `axis` functions using the owl data.

7.5.3 Adding Extra Points, Text, and Lines

This section addresses features that can be used to increase the visual appeal of graphs. Possible embellishments might be different types of lines and points, grids, legends, transformed axes, and much more. Look at the `par` help file, obtained by typing ?par, to see many of the features that can be added and altered. We could write an entire book on the `par` options, some of which have been addressed in Chapter 5 and in earlier sections of this chapter. More are discussed in Chapter 8. However, even novice users will feel the need for some information on the `par` function at an early point. Because we do not want this volume to become phonebook-sized, we discuss some of the `par` and plotting options in a birds-eye overview mode, and try to guide you to the appropriate help files.

The functions `points`, `text`, and `lines` are valuable companions when working in R and were used in some earlier chapters.

The function `points` adds new values to a plot, such as x-values and (option-ally) y-values. By default, the function plots points, so, just as with `plot`, `type` is set to "p". However, all the other types can be used: "l" for lines, "o" for overplotted points and lines, "b" for points and lines, "s" and "S" for steps, and "h" for vertical lines. Finally, "n" produces a graph-setup with no data points or lines (see Section 7.5.2). Symbols can be changed using `pch` (see Chapter 5).

The function text is similar to points in that it uses x and (optionally) y-coordinates but adds a vector called labels containing the label strings to be positioned on the graph. It includes extra tools for fine-tuning the placement of the string on the graph, for example, the attributes pos and offset. The pos attribute indicates the positions below, to the left of, above, and to the right of the specified coordinates (respectively, 1, 2, 3, 4) and offset gives the offset of the label from the specified coordinate in fractions of a character width. These two options become relevant with long character strings that are not displayed properly in R's default display.

We have seen the lines function in Chapter 5. It is a function that accepts coordinates and joins the corresponding points with lines.

7.5.4 Using type = "n"

With the plot function, it is possible to include the attribute type = "n" to draw everything but the data. The graph is set up for data, including axes and their labels. To exclude these, add axes = FALSE, xlab = "", ylab = "". It then appears there is nothing left. However, this is not the case, because the plot retains the data that were entered in the first part of the plot function. The user is now in full control of constructing the plot. Do you want axes lines? If so, where do you want them and how do you want them to look? Do you want to display the data as points or as lines? Everything that is included in the default plot, and much more, can be altered and added to your plot. Here are some of the available variations:

Command	Description
abline	Adds an a,b (intercept, slope) line, mainly regression, but also vertical and horizontal lines
arrows	Adds arrows and modifies the head styles
Axis	Generic function to add an axis to a plot
axis	Adds axes lines
box	Adds different style boxes
contour	Creates a contour plot, or adds contour lines to an existing plot
curve	Draws a curve corresponding to the given function or expression
grid	Adds grid to a plot
legend	Adds legend to a plot
lines	Adds lines
mtext	Inserts text into the margins of the figure or in the margin of the plot device
points	Adds points, but may include type command
polygon	Draws polygons with vertices defined by x and y
rect	Draws rectangles
rug	Adds a one dimensional representation of the data to the plot on one of the two axes.
Segments	Adds line segments
text	Adds text inside the plot
title	Adds a title

7.5.5 Legends

The function legend appears difficult at first encounter, but is easily mastered. In Fig. 7.9, a legend was added to a Cleveland dotplot. The code is

```
> legend("bottomright", c("values", "mean"),
    pch = c(1, 19), bg ="white")
```

The first attribute may consist of an *x*- and *y*-coordinate, or an expression such as shown here. Other valid expressions are "bottom", "bottom-left", "left", "topleft", "top", "topright", "right", and "center". Consult the legend help file for more options.

Zuur et al. (2009) used a bird dataset that was originally analysed in Loyn (1987), and again in Quinn and Keough (2002). Forest bird densities were measured in 56 forest patches in southeastern Victoria, Australia. The aim of the study was to relate bird densities to six habitat variables: (1) size of the forest patch, (2) distance to the nearest patch, (3) distance to the nearest larger patch, (4) mean altitude of the patch, (5) year of isolation by clearing, and (6) an index of stock grazing history (1 = light, 5 = intensive). A detailed analysis of these data using linear regression is presented in Appendix A of Zuur et al. (2009). The optimal linear regression model contained LOGAREA and GRAZE (categorical). To visualise what this model is doing, we plot the fitted values. There are five grazing levels, and, therefore, the linear regression (see the summary command below) gives an equation relating bird abundance to LOGAREA for each grazing level. These are given by

Observations with GRAZE = 1:	$ABUND_i = 15.7 + 7.2 \times LOGAREA_i$
Observations with GRAZE = 2:	$ABUND_i = 16.1 + 7.2 \times LOGAREA_i$
Observations with GRAZE = 3:	$ABUND_i = 15.5 + 7.2 \times LOGAREA_i$
Observations with GRAZE = 4:	$ABUND_i = 14.1 + 7.2 \times LOGAREA_i$
Observations with GRAZE = 5:	$ABUND_i = 3.8 + 7.2 \times LOGAREA_i$

Readers familiar with linear regression will recognise this as a linear regression model in which the intercept is corrected for the levels of the categorical variable. Next, we (i) plot the ABUNDANCE data versus LOGAREA, (ii) calculate fitted values for the five grazing regimes, (iii) add the five lines, and (iv) add a legend. The resulting graph is presented in Fig. 7.11. The following shows step by step how it was created.

First, read the data, apply the log transformation, and use the plot function. We have previously used similar R code:

```
> setwd("C:/RBook/")
> Birds <- read.table(file = "loyn.txt", header = TRUE)
> Birds$LOGAREA <- log10(Birds$AREA)
```

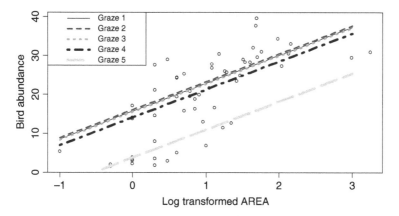

Fig. 7.11 Five fitted lines for the Loyn bird data. Each line is for a particular grazing regime

```
> plot(x = Birds$LOGAREA, y = Birds$ABUND,
      xlab = "Log transformed AREA",
      ylab = "Bird abundance")1
```

To see the source of the five slopes and the intercept, use the code:

```
> M0 <- lm(ABUND~ LOGAREA + fGRAZE, data = Birds)
> summary(M0)
```

If you are not familiar with linear regression, do not spend time struggling to comprehend this. The summary output contains the required information. To predict fitted bird abundances per grazing level, we need the LOGAREA values. The simplest method is to look at Fig. 7.11 and choose several arbitrary values within the range of the observed data, say –1, 0, 1, 2, and 3:

```
> LAR <- seq(from = -1, to = 3, by = 1)
> LAR
```

```
[1] -1 0 1 2 3
```

Now we determine the abundance values per grazing level using simple calculus and R code:

```
> ABUND1 <- 15.7 + 7.2 * LAR
> ABUND2 <- 16.1 + 7.2 * LAR
> ABUND3 <- 15.5 + 7.2 * LAR
> ABUND4 <- 14.1 + 7.2 * LAR
> ABUND5 <- 3.8 + 7.2  * LAR
```

Adding the fitted values as lines to the graph is also familiar territory (see Chapter 5). We do not have a spaghetti problem, as the AREA data are sorted from – 1 to 3.

```
> lines(LAR, ABUND1, lty = 1, lwd = 1, col =1)
> lines(LAR, ABUND2, lty = 2, lwd = 2, col =2)
> lines(LAR, ABUND3, lty = 3, lwd = 3, col =3)
> lines(LAR, ABUND4, lty = 4, lwd = 4, col =4)
> lines(LAR, ABUND5, lty = 5, lwd = 5, col =5)
```

We added visual interest with different line types, widths, and colours. Finally, it is time to add the legend; see the R code below. First we define a string legend.txt with five values containing the text that we want to use in the legend. The legend function then places the legend in the top left position, the line in the legend for the first grazing level is black (col = 1), solid (lty = 1), and has normal line width (lwd = 1). The line in the legend for grazing level 5 is light blue (col = 5), has the form --- (lty = 5) and is thick (lwd = 5).

```
> legend.txt <- c("Graze 1", "Graze 2",
                  "Graze 3", "Graze 4", "Graze 5")
> legend("topleft", legend = legend.txt,
        col = c(1, 2, 3, 4, 5),
        lty = c(1, 2, 3, 4, 5),
        lwd = c(1, 2, 3, 4, 5),
        bty = "o", cex = 0.8)
```

The attribute cex specifies the size of the text in the legend, and the bty adds a box around the legend.

 Do Exercise 8 in Section 7.10. In this exercise, smoothers are used for the male and female owl data and are superimposed onto the graph. The legend function is used to identify them.

7.5.6 Identifying Points

The function identify is used to identify (and plot) points on a plot. It can be done by giving the *x*, *y* coordinates of the plot or by simply entering the plot object (which generally defines or includes coordinates). Here is an example:

```
> plot(y = Benthic$Richness, x = Benthic$NAP,
      xlab = "Mean high tide (m)",
      ylab = "Species richness", main = "Benthic data")
> identify(y = Benthic$Richness, x = Benthic$NAP)
```

With the attribute labels in the identify function, a character vector giving labels for the points can be included. To specify the position and offset of the labels relative to the points; place your mouse near a point and left-click; R will plot the label number close to the point. Press "escape" to cancel the process. It is also possible to use the identify function to obtain the sample numbers of certain points; see its help file. Note that the identify function only works for graphs created with the plot function, and not with boxplots, dotcharts, bar charts, pie charts, and others.

7.5.7 Changing Fonts and Font Size*

This section is a bit more specialised and may be skipped upon first reading. Fonts and font sizes are somewhat peculiar in R. When you open a graphing device you can apply an attribute pointsize that will be the default point size of plotted text. Default font mappings are provided for four device-independent font family names: "sans" for a sans-serif font, "serif" for a serif font, "mono" for a monospaced font, and "symbol" for a symbol font. Type windowsFonts() to see the font types that are currently installed.

Font defines the font face. It is an integer that specifies which font face to use for text. If possible, device drivers are organized so that 1 corresponds to plain text, 2 to bold face, 3 to italic, and 4 to bold italic. To modify the default font, we usually draw plots omitting the component for which we want to change the default font and code it separately, including options for font size, font face, and font family. For example, to add a title in a serif font to Fig. 7.11, use

```
> title("Bird abundance", cex.main = 2,
        family = "serif", font.main = 1)
```

This would plot "Bird abundance" as a title twice the default size, with a serif font style in normal font face. For title there are special options for font size and font face, cex.main and font.main. Sometimes you may need to specify the family using par. You can also change font and size for text, mtext, axis, xlab, and ylab. Consult the help file for par for specific information on changing fonts.

7.5.8 Adding Special Characters

Often you may want to include special characters in legends or labels. This is not difficult in R, although it may require searching in several help files to find exactly what you want. The function that is mostly used is expression. You can get an impression of the possibilities by typing demo (plotmath).

Here is a brief example: Mendes et al. (2007) measured the nitrogen isotopic composition in growth layers of teeth from 11 sperm whales stranded in Scotland. Figure 7.12 shows a scatterplot of nitrogen isotope ratios versus age, for one particular whale, nicknamed Moby. The y-label of the graph contains the expression $\delta^{15}N$. It is tempting to import this graph without the y-label into Word and add the $\delta^{15}N$ before submission to a journal, but it can easily be done in R using this code:

```
> setwd("C:/RBook/")
> Whales <- read.table(file="TeethNitrogen.txt",
               header = TRUE)
> N.Moby <- Whales$X15N[Whales$Tooth == "Moby"]
> Age.Moby <- Whales$Age[Whales$Tooth == "Moby"]
> plot(x = Age.Moby, y = N.Moby, xlab = "Age",
       ylab = expression(paste(delta^{15}, "N")))
```

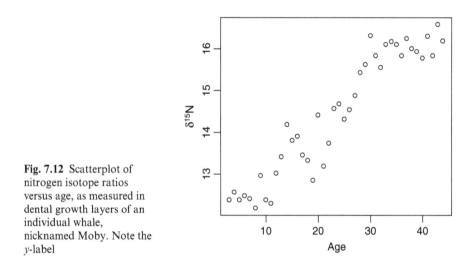

Fig. 7.12 Scatterplot of nitrogen isotope ratios versus age, as measured in dental growth layers of an individual whale, nicknamed Moby. Note the y-label

The paste command joins the δ^{15} and N, and the expression function inserts the $\delta^{15}N$.

7.5.9 Other Useful Functions

There are a number of other functions that may come in handy when making graphs. Consult the help files for attributes that may, or must, be provided.

Function[a]	Description
plot.new	Opens a new graphics frame, same as frame()
win.graph	Opens extra second graph window. You can set width and height of the screen
windows	Similar to win.graph but with more options
savePlot	Saves current plot as ("wmf", "emf", "png", "jpeg", "jpg", "bmp", "ps", "eps", or "pdf")
locator	Records the position of the cursor by clicking left cursor; stops by clicking right cursor
range	Returns a vector containing the minimum and maximum of all the given arguments; useful for setting x or y limits
matplot	Plots columns of one matrix against the columns of another; especially useful when multiple Y columns and a single X. See also matlines and matpoints for adding lines and points, respectively
persp	Perspective plots of surfaces over an x–y plane
cut	Converts a numeric variable into a factor
split	Divides a vector or data frame with numeric values into groups

[a]Don't forget to include the brackets with these functions!

7.6 The Pairplot

In the previous graph, we used the plot function to make a scatterplot of two continuous variables; the following demonstrates scatterplots for multiple continuous variables. This could be done by using the plot function to plot variable 1 versus 2, 1 versus 3, 1 versus 4, and so on, and following with mfrow and mar to put it all into a single graph. However, the R function pairs can be used to produce a multipanel scatterplot. We use the benthic data for illustration:

```
> setwd("C:/RBook/")
> Benthic <- read.table(file = "RIKZ2.txt",
                        header = TRUE)
> pairs(Benthic[, 2:9])
```

The first two lines import the data, and the pairs function is applied to all the variables from the data frame Benthic with the exception of the first column, which contains the labels. The resulting graph is presented in Fig. 7.13[2].

We have included species richness as the first variable. As a result, the first row of the plot contains graphs of all variables against richness. The rest of the plot shows graphs of all variables versus one another. From a statistical point of view, we want to model richness as a function of all the other variables, hence

---

[2] Using the command plot (Benthic [, 2:9]) will give the same graph, because Benthic is a data frame, and the plot function recognises this and calls the function plot.data.frame.

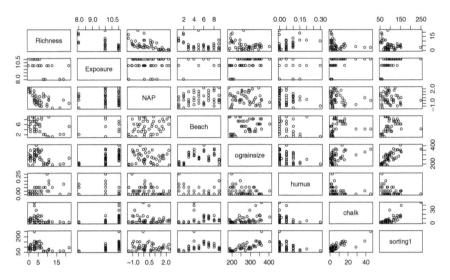

Fig. 7.13 Scatterplot matrix for variables in the benthic data. The diagonal shows the name of the variable which is on the *x*-axis below and above it, and on the *y*-axis left and right of it

clear relationships in the first row (or column) are good, whereas clear patterns in the other panels (collinearity) are not good at all. The pairplot shows clear relationships between some of the variables, for example, between species richness and NAP and between grain size and sorting (this makes biological sense, as sorting is a measure of energy).

7.6.1 Panel Functions

Half of the information in the pairplot appears superfluous, in as much as every graph appears twice, once above the diagonal and once below, but with the axes reversed. It is possible to specify panel functions to be applied to all panels, to the diagonal panels, or to the panels above or below the diagonal (Fig. 7.14). The R code for this can be found at the end of the `pairs` help file obtained by entering `? pairs` into the R console window.

```
> pairs(Benthic[, 2:9], diag.panel = panel.hist,
        upper.panel = panel.smooth,
        lower.panel = panel.cor)
Error in pairs.default(Benthic[, 2:9], diag.panel =
panel.hist, upper.panel = panel.smooth,: object
"panel.cor" not found
```

The problem here is that R does not recognise the `panel.cor` and the `panel.hist` functions. These specific pieces of code from the end of the

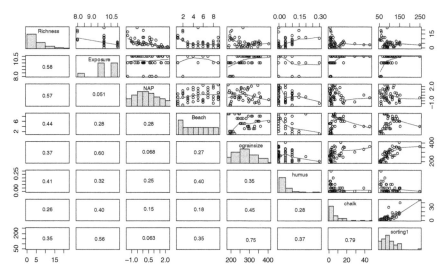

Fig. 7.14 The extended pairplot using histograms on the diagonal, scatter plots with smoothers above the diagonal, and Pearson correlation coefficients with size proportionate to the correlation below the diagonal. The code was taken from the `pairs` help file

`pairs` help file must be copied and pasted into the R console. Copy the entire function and rerun the `pairs` command above. For specific advice, see the online R code for this book, which can be found at www.highstat.com. The `panel.cor` and `panel.hist` code is complicated and beyond the scope of this book, so is not addressed here. Simply copy and paste it.

If you are interested in using Pearson correlation coefficients in a pairplot, see http://www.statmethods.net/graphs/scatterplot.html. This provides an example, as well as a link to the package and a function that can be used to colour entire blocks based on the value of the Pearson correlation.

 Do Exercise 9 in Section 7.10. In this exercise, the `pairs` function is used for the vegetation data.

7.7 The Coplot

7.7.1 A Coplot with a Single Conditioning Variable

The `pairs` function shows only two-way relationships. The next plotting tools we discuss can illustrate three-way, or even four-way, relationships. This type of plot is called a conditioning plot or coplot and is especially well suited to visualizing how a response variable depends on a predictor, given other

Fig. 7.15 Coplot for the benthic data. The *lower left* panel represents a scatter!plot of richness versus NAP for beach 1, the *lower right* panel for beach 3, the *middle left* for beach 4, and the *upper right* for beach 9

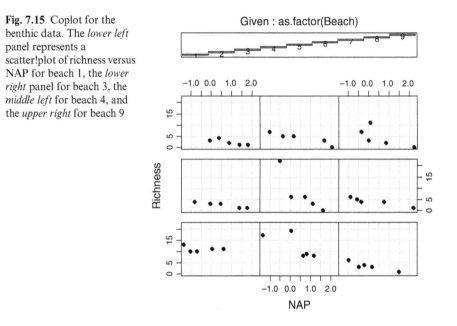

predictors. Figure 7.15 is a plot of the RIKZ data using the variables Beach, NAP, and Richness. The nine graphs represent beaches one to nine, which are listed at the top and displayed in the separate panels, called the dependence panels. Starting at the bottom row and going from left to right, the first row depicts beaches one to three, the second row four to six, and the top row beaches seven to nine. As you can see, the beach numbers are also given, although not well placed. The R code to make the graph in Fig. 7.15 is as follows.

```
> setwd("C:/RBook/")
> Benthic <- read.table(file = "RIKZ2.txt",
                        header = TRUE)
> coplot(Richness ~ NAP | as.factor(Beach), pch=19,
        data = Benthic)
```

The function `coplot` uses a different notation than the `plot` function. Variables to be plotted are given in a formula notation that uses the tilde operator $\sim$ as a separator between the dependent and the independent variables. Contrary to what you have been using in the `plot` function where the first variable is assumed to be the *x*-variable and the second variable the *y*-variable, the formula notation always uses $y \sim x$. The above code thus directs R to plot species richness (R) versus NAP. The addition of | `as.factor(Beach)` creates the panels and indicates that the plot should be produced conditional on the variable Beach, which is first coerced into a factor. The `data` attribute gives the command to look in the `Benthic` data frame for the variables used in the formula.

Instead of using a categorical variable for the conditioning variable, we can use a continuous variable, for example, grainsize. The following code creates Fig. 7.16. Scatterplots of richness versus NAP are drawn for different grainsize values.

```
> coplot(Richness ~ NAP | grainsize, pch=19,
          data = Benthic)
```

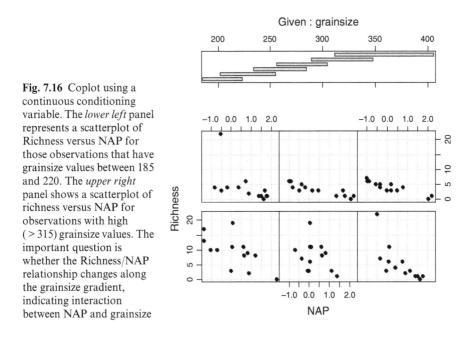

Fig. 7.16 Coplot using a continuous conditioning variable. The *lower left* panel represents a scatterplot of Richness versus NAP for those observations that have grainsize values between 185 and 220. The *upper right* panel shows a scatterplot of richness versus NAP for observations with high (> 315) grainsize values. The important question is whether the Richness/NAP relationship changes along the grainsize gradient, indicating interaction between NAP and grainsize

The grainsize values were divided into six overlapping groups with approximately equal numbers of points. If the Richness/NAP relationship changes along the grainsize gradient, giving a visual indication of the presence of an interaction between NAP and grainsize, it may be worthwhile to include this interaction term in, for example, a linear regression model.

The coplot function contains a large number of arguments that can be used to create exciting plots. See its help file, obtained by ? coplot. The most useful is panel, which takes a function that is carried out in each panel of the display. By default coplot uses the points function, but we can easily create our own function and apply it to each panel. For example, we may wish to add a linear regression line to each panel in Fig. 7.15 (Fig. 7.17). If all the lines turn out to be parallel, there is no visual evidence of an interaction between beach and NAP (i.e., the richness ~ NAP relationship is the same along the entire stretch of coastline). In this case, the lines do differ. Here is the code that created Fig. 7.17:

Fig. 7.17 Coplot of the
RIKZ data showing species
richness versus NAP with a
separate panel for each of
the nine beaches

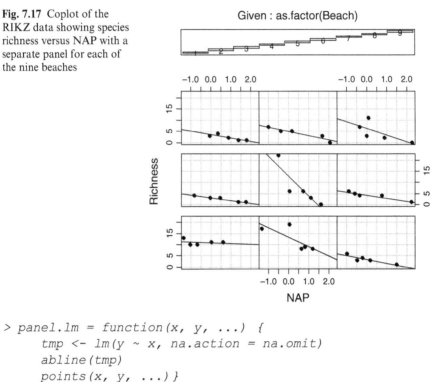

```
> panel.lm = function(x, y, ...) {
    tmp <- lm(y ~ x, na.action = na.omit)
    abline(tmp)
    points(x, y, ...)}
> coplot(Richness ~ NAP | as.factor(Beach), pch = 19,
        panel = panel.lm, data = Benthic)
```

The function `panel.lm` defines how the data should be displayed in each
panel. Three dots at the end indicate that other arguments may be supplied that
will be evaluated in the function. The linear regression function `lm` is used to
store the data temporarily in the variable `tmp`, and any NAs are omitted from
the analysis. The function `abline` plots the line, and the function `points`
plots the points.

Another predefined `panel` function is `panel.smooth`. This uses the
LOESS smoother to add a smooth line.

As you can see above, we defined our own `panel` function. This facility is
useful for creating customized panel functions for use with `coplot`. For
example, means and confidence limits can be added to each panel, and confidence limits can be added to regression lines.

`Coplot` is also a good tool for investigating the amount of data in each
combination of covariates.

Do Exercise 10 in Section 7.10. This exercise creates a coplot of the
vegetation data.

7.7.2 The Coplot with Two Conditioning Variables

One can include a third predictor variable in a coplot, but the benthic data do not yield much additional information when one of the other variables is included. Therefore we present another example: a subset of data analysed in Cruikshanks et al. (2006). The data are available in the file *SDI2003.txt*. The original research sampled 257 rivers in Ireland during 2002 and 2003. One of the aims was to develop a new tool for identifying acid-sensitive waters, which is currently done by measuring pH levels. The problem with pH is that it is extremely variable within a catchment and depends on both flow conditions and underlying geology. As an alternative measure, the Sodium Dominance Index (SDI) was proposed. Of the 257 sites, 192 were nonforested and 65 were forested. Zuur et al. (2009) modelled pH as a function of SDI, forested or nonforested, and altitude, using regression models with spatial correlation.

The relationship between pH and SDI may have been affected by the altitude gradient and forestation. Calculating this demands a three-way interaction term between two continuous (SDI and altitude) and one categorical (forestation) explanatory variable. Before including such an interaction in a model, we can visualise the relationships with the coplot. In the previous section, we used coplots with a single conditioning variable; here we use two conditioning variables. We use the log-transformed altitude values. The coplot is shown in Fig. 7.18. The R code is as follows.

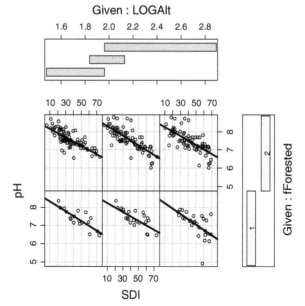

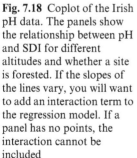

Fig. 7.18 Coplot of the Irish pH data. The panels show the relationship between pH and SDI for different altitudes and whether a site is forested. If the slopes of the lines vary, you will want to add an interaction term to the regression model. If a panel has no points, the interaction cannot be included

```
> setwd("C:/RBook/")
> pHEire <- read.table(file = "SDI2003.txt",
                            header = TRUE)
> pHEire$LOGAlt <- log10(pHEire$Altitude)
> pHEire$fForested <- factor(pHEire$Forested)
> coplot(pH ~ SDI | LOGAlt * fForested,
        panel = panel.lm, data = pHEire)
```

We use the same `panel.lm` function as in the previous section. (This requires copying and pasting it into the R console, if R has been shut down.) Because the variable `LOGAlt`, the logarithmically transformed altitude, is numeric, it is divided into a number of conditioning intervals, and, for each interval, pH is plotted against SDI. In addition, the data are segregated based on the Forested factor. The number and position of intervals for `LOGAlt` can be controlled with the `given.values` argument; see the `coplot` help file. Without this argument, the numeric variable is divided into six intervals overlapping by approximately 50. An easier approach may be using the `number` argument. Run this command:

```
> coplot(pH ~ SDI | LOGAlt * fForested,
        panel = panel.lm, data = pHEire, number = 2)
```

Compare the resulting coplot (which is not shown here) with that in Fig. 7.18; this one has fewer panels. The `number` argument can also be used if the coplot crashes due to an excessive number of panels.

7.7.3 Jazzing Up the Coplot*

This section is slightly more complicated (hence the asterisk in the title), and may be omitted upon first reading.

Figure 7.18 shows the relationship between pH versus SDI, altitude and forestation (and their interactions). To demonstrate what can be done, we produce the same coplot as that in Fig. 7.18, but with points of different colours depending on temperature. Temperatures above average are indicated by a light grey dot, and those below average are shown by a dark dot (obviously, red and blue dots would be better). Before this can be done, we need to use the following code to create a new variable containing the grey colours.

```
> pHEire$Temp2 <- cut(pHEire$Temperature, breaks = 2)
> pHEire$Temp2.num <- as.numeric(pHEire$Temp2)
```

The `cut` function separates the temperature data into two regimes, because we use `breaks = 2`. We encounter a problem in that the output, Temp2, is a factor, as can be seen from entering:

```
> cut(pHEire$Temperature, breaks = 2)
  [1] (1.89,7.4]  (1.89,7.4]  (1.89,7.4]  (1.89,7.4]
  [5] (1.89,7.4]  (1.89,7.4]  (1.89,7.4]  (1.89,7.4]
  [9] (1.89,7.4]  (1.89,7.4]  (1.89,7.4]  (1.89,7.4]
 [13] (1.89,7.4]  (1.89,7.4]  (7.4,12.9]  (1.89,7.4]
  ...
[197] (7.4,12.9]  (7.4,12.9]  (7.4,12.9]  (7.4,12.9]
[201] (7.4,12.9]  (7.4,12.9]  (7.4,12.9]  (7.4,12.9]
[205] (7.4,12.9]
Levels: (1.89,7.4] (7.4,12.9]
```

Each temperature value is allocated to either the class 1.89 − 7.4 (below average) or 7.4 − 12.9 (above average) degrees Celsius. A factor cannot be used for colours or greyscales; therefore we convert Temp2 to a number, using the as.numeric function. As a result, pHEire$Temp2.num is a vector with values 1 and 2. We could have done this in Excel, but the cut function is more efficient. We are now ready to create the coplot in Fig. 7.19, using the following R code.

```
> coplot(pH ~ SDI | LOGAlt * fForested,
    panel = panel.lm, data = pHEire,
    number = 3, cex = 1.5, pch = 19,
    col = gray(pHEire$Temp2.num / 3))
```

It seems that high pH values were obtained for low SDI values with Forested = 2 (2 represents nonforested and 1 is forested) and above average temperature.

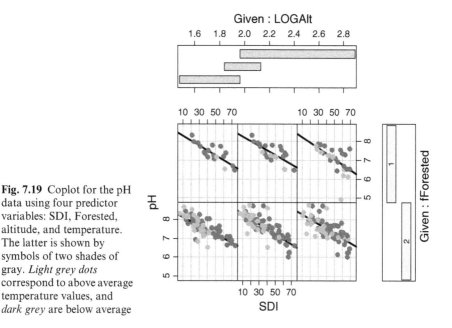

Fig. 7.19 Coplot for the pH data using four predictor variables: SDI, Forested, altitude, and temperature. The latter is shown by symbols of two shades of gray. *Light grey dots* correspond to above average temperature values, and *dark grey* are below average

7.8 Combining Types of Plots*

Here we touch upon R's more advanced graphing possibilities. There are several graphing systems that can be used in R. All the graphs we have shown were made by using the base package *graphics*. The R package called *grid* offers many advanced possibilities. It is possible to combine different plots into a single graph. We have already used the mfrow command to enable plotting several graphs on one screen. Here we use layout to create complex plot arrangements. Figure 7.20 shows a scatterplot of species richness versus NAP and also includes the boxplots of each variable.

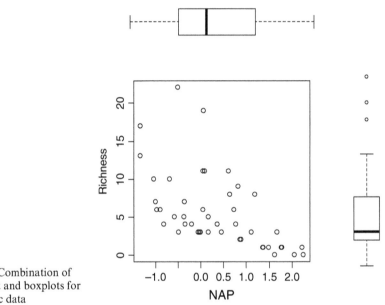

Fig. 7.20 Combination of scatterplot and boxplots for the benthic data

To produce this graph, we first need to define the number of graphs to incorporate, their placement, and their size. In this case, we want to arrange a 2-by-2 window with the scatterplot in the lower left panel, one of the boxplots in the upper left panel, and one boxplot in the lower right panel. For this we define a matrix, let's call it MyLayOut, with the following values.

```
> MyLayOut <- matrix(c(2, 0, 1, 3), nrow = 2, ncol=2,
                      byrow = TRUE)
> MyLayOut
      [,1] [,2]
[1,]    2    0
[2,]    1    3
```

The matrix command was introduced in Chapter 2. It looks intimidating, but simply creates a matrix with the elements 2 and 0 on the first row, and 1 and 3 on the second row. We use this matrix inside the layout function, followed by three plot commands. The first graph appears in the lower left corner (specified by the 1 in the matrix), the second plot in the upper left (specified by the 2), and the third graph in the lower right. Because there is a 0 in the upper right position of MyLayout, no graph will be drawn in that quadrant.

The next part of the code consists of

```
> nf <- layout(mat = MyLayOut, widths = c(3, 1),
          heights = c(1, 3), respect = TRUE)
```

The widths option specifies the relative width of the columns. In this case, the first column, containing the scatterplot and the boxplot for NAP, is 3, and the second column, containing the boxplot for richness, has a width of 1. The heights column specifies the height of the rows. The respect = TRUE ensures that a 1-unit in the vertical direction is the same as a 1-unit in the horizontal direction. The effect of these settings in the layout function can be visualised with the following command.

```
> layout.show(nf)
```

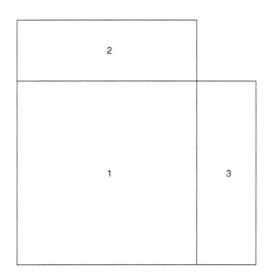

Fig. 7.21 Layout of the graphical window. The results of the first plot command will go into panel 1 (*lower left*), the next into panel 2, and the third plot into panel 3

All that remains is to make the three graphs. We must ensure that the range of the boxplot in panel 2 is synchronised with the range of the horizontal axis in panel 1, and the same holds for panel 3 and the vertical axis in panel 1. We also need to avoid excessive white space around the graphs, which means some trial and error with the `mar` values for each graph. We came up with the following code.

```
> xrange <- c(min(Benthic$NAP), max(Benthic$NAP))
> yrange <- c(min(Benthic$Richness),
              max(Benthic$Richness))
> #First graph
> par(mar = c(4, 4, 2, 2))
> plot(Benthic$NAP, Benthic$Richness, xlim = xrange,
       ylim = yrange, xlab = "NAP", ylab = "Richness")
> #Second graph
> par(mar = c(0, 3, 1, 1))
> boxplot(Benthic$NAP, horizontal = TRUE, axes = FALSE,
       frame.plot = FALSE, ylim = xrange, space = 0)
> #Third graph
> par(mar = c(3, 0, 1, 1))
> boxplot(Benthic$Richness, axes = FALSE,
          ylim = yrange, space = 0, horiz = TRUE)
```

Most of the options are self-explanatory. Change the values of the `mar`, and see what happens. Another function that can be used for similar purposes is the `split.screen`; see its help file.

7.9 Which R Functions Did We Learn?

Table 7.1 shows the R functions that were introduced in this chapter.

Table 7.1 R functions introduced in this chapter

Function	Purpose	Example
pie	Makes a pie chart	pie(x)
pie3D	Makes a 3-D piechart	pie3D(x)
par	Sets graph parameters	par(...)
barplot	Makes a bar chart	barplot(x)
arrows	Draws arrows	arrows(x1,y1,x2,y2)
box	Draws a box around the graph	box()
boxplot	Makes a boxplot	boxplot(y)
		boxplot(y~x)
text	Adds text to a graph	text(x,y,"hello")
points	Adds points to an existing graph	points(x,y)
legend	Adds a legend	legend("topleft", MyText,
		lty = c(1,2,3))

Table 7.1 (continued)

Function	Purpose	Example
title	Adds a title	title(MyText)
expression	Allows for special symbols	ylab = expression(paste(deltao{15}, "N"))
pairs	Creates multipanel scatterplots	Pairs(X)
coplot	Creates multipanel scatterplots	Coplot(y~x|z)
layout	Allows for multiple graphs in the same window	layout(mat,widths,heights) plot(x) plot(y)

7.10 Exercises

Exercise 1. The use of the `pie` function using the avian influenza data.
In Section 7.1, we used the total number of bird flu cases per year. Make a pie chart to illustrate the totals by country. Place the labels such that they are readable. The file *BirdFludeaths.txt* contains the data on deaths from the disease. Make a pie chart showing total deaths per year and one showing deaths per country.

Exercise 2. The use of the `barchart` and `stripchart` functions using a vegetation dataset.
In Section 4.1, we calculated species richness, as well as its mean values and standard deviations, in eight transects. Make a bar chart for the eight mean values and add a vertical line for the standard error.

Make a graph in which the means are plotted as black points, the standard errors as lines around the mean, and the observed data as open dots.

Exercise 3. The use of the `boxplot` function using a vegetation dataset.
Using the vegetation data in Exercise 2, make a boxplot showing the richness values.

Exercise 4. The use of the `boxplot` function using a parasite dataset.
In Section 6.3.3, a cod parasite dataset was used. Make a boxplot of the number of parasites (Intensity) conditional on area, sex, stage, or year. Try combinations to detect interactions.

Exercise 5. The use of the `dotchart` function using the owl data.
In Section 7.3, we used the owl data. Make two Cleveland dotplots of nestling negotiation and arrival time. Make a Cleveland dotplot showing arrival time per night. The nest and food treatment variables show which observations were made on the same night. See also Exercise 2 in Section 6.6.

Exercise 6. The use of the `dotchart` function using the parasite data.
Make a Cleveland dotplot for the parasite data that were used in Exercise 4. Use the number of parasites (Intensity), and group the observations by area,

sex, stage, or by year. Make a Cleveland dotplot showing depth, and group the observations by prevalence.

Exercise 7. The use of the `plot` and `axis` functions using the owl data.

Apply a logarithmic transformation (use 10 as the base) on the nestling negotiation data. Add the value of 1 to avoid problems with the log of 0. Plot the transformed nestling negotiation data versus arrival time. Note that arrival time is coded as 23.00, 24.00, 25.00, 26.00, and so on. Instead of using the labels 25, 26, etc. for arrival time, use 01.00, 02.00, and so on.

Make the same graph, but use back-transformed values as labels along the vertical axis. This means using the log-transformed nestling negotiation data but with the label 1 if the log-transformed value is 0, 10 if the log-transformed value is 1, and so on.

Exercise 8. The use of the `legend` function using the owl data.

Add a smoother (see Chapter 5) to the graph created in Exercise 7 to visualise the pattern for the male data and for the female data. Extract the data from the males, fit a smoother, and superimpose this line onto the graph. Do the same for the female data. Use a legend to identify the different curves. Do the same for food treatment and night.

Exercise 9. The use of the `pairs` function using the vegetation data.

Make a pairplot for all the climatic variables in the vegetation data. Add correlation coefficients in the lower panels. What does the graph tell you?

Exercise 10. The use of the `coplot` function using the vegetation data.

Plot species richness versus a covariate of your choice conditional on transect.

Chapter 8
An Introduction to the Lattice Package

R contains many functions that can be used to draw graphs for specific types of data. Many of these plotting functions are not available in the packages that are loaded by default when R is started. Some are as simple to use as the `plot` function; others require more effort. The lattice package allows R to reach its full potential for imaging higher-dimensional data. We used one type of lattice plot, the multipanel scatterplot, in Section 1.4.1 to plot density of deep-sea pelagic bioluminescent organisms versus depth.

The lattice package was written by Deepayan Sarkar, who recently published an excellent book which we highly recommend (Sarkar, 2008). The package implements the Trellis Graphics framework developed at Bell Labs in the early 1990s.

In Chapter 7 we introduced the `coplot` function, which is particularly useful for displaying subsets of data in separate panels, when there is a grouping structure to the data. The lattice package allows taking this feature much further, but it comes at the price of more programming effort. However, by now you should have gained enough proficiency to master the function without too much difficulty.

8.1 High-Level Lattice Functions

The lattice user interface primarily consists of a number of generic functions called "high-level" functions, each designed to create a particular type of statistical display (Table 8.1). Fortunately, it is not necessary to learn each function individually. They are designed with a similar formula interface for different types of multipanel conditioning and respond to a large number of common arguments. Hence, once one function is mastered, learning to use the other functions is simple.

The plotting is performed by a default panel function embedded in each generic function that is applied to each panel. Most often the user will not be aware that a panel function is responding to arguments given in the function call. Names of default panel functions are generally self-explanatory. For example, the default

A.F. Zuur et al., *A Beginner's Guide to R*, Use R,
DOI 10.1007/978-0-387-93837-0_8, © Springer Science+Business Media, LLC 2009

panel function for the high-level function `histogram` is `panel.histogram`, for `densityplot` it is `panel.densityplot`, and for `xyplot` it is `panel.xyplot`, and so on. These predefined functions are available for you to use and to modify. We discuss panel functions more fully in Section 8.6.

Table 8.1 shows some of the high-level functions available in lattice.

Table 8.1 High-level functions of the lattice package

Function	Default Display
`histogram()`	Histogram
`densityplot()`	Kernel density plot
`qqmath()`	Theoretical quantile plot
`qq()`	Two-sample quantile plot
`stripplot()`	Stripchart (comparative 1-D scatterplots)
`bwplot()`	Comparative box-and-whisker plots
`dotplot()`	Cleveland dotplot
`barchart()`	Barplot
`xyplot()`	Scatterplot
`splom()`	Scatterplot matrix
`contourplot()`	Contour plot of surfaces
`levelplot()`	False colour level plot of surfaces
`wireframe()`	Three-dimensional perspective plot of surfaces
`cloud()`	Three-dimensional scatterplot
`parallel()`	Parallel coordinates plot

 Do Exercise 1 in Section 8.11. This introduces lattice plots and provides an overview of the possibilities of the package.

8.2 Multipanel Scatterplots: `xyplot`

In the exercises in Chapter 4 we used temperature data measured at 30 stations along the Dutch coastline over a period of 15 years. Sampling took place 0 to 4 times per month, depending on the season. In addition to temperature, salinity was recorded at the same stations, and these measurements are used here. The data (in the file *RIKZENV.txt*) are submitted to the `xyplot` function to generate a multipanel scatterplot. The following is the code to enter the data into R, create a new variable, `MyTime`, representing time (in days), and create a multipanel scatterplot.

```
> setwd("C:/RBook")
> Env <- read.table(file ="RIKZENV.txt", header = TRUE)
```

```
>Env$MyTime <- Env$Year + Env$dDay3 / 365
>library(lattice)

>xyplot(SAL ~ MyTime | factor(Station), type = "l",
     strip = function(bg, ...)
       strip.default(bg = 'white', ...),
 col.line = 1, data = Env)
```

The function `xyplot` contains characteristics common to all high-level
lattice functions. The most obvious ones are the use of a formula, the vertical
bar (also called the pipe symbol) within the formula, and the `data` argument.
Lattice uses a formulalike structure that is also used in R for statistical models.
The variable preceding the tilde (the ~) is plotted on the *y*-axis with that
following along the *x*-axis. The conditioning variable (in this case `Station`),
which will generate multiple panels, follows the vertical bar.

When there is no conditioning variable, the result of `xyplot` will be similar
to the normal `plot` function; the data will be plotted in a single panel. The
conditioning variable is usually a factor (note that we typed: `factor(Sta-
tion)`), but it may also be a continuous variable. The default behaviour when
using a continuous variable for conditioning is to treat each of its unique values
as a discrete level. Often, however, the variable may contain so many values that
it is advisable to partition it into intervals. This can be achieved by using the
functions `shingle` and `equal.count`; see their help pages.

Figure 8.1 displays the data in five rows of 6 panels, showing a graph for each
station. The station name is given in the horizontal bar, called the strip, above
the panel.

The code is not difficult. The graph is drawn by the `xyplot` function using a
formula to plot salinity versus (~) time, conditional on (|) station. We added two
`xyplot` arguments: `strip`, used to create a white background in each strip, and
`col.line` = 1 to designate black lines (recall from Chapter 5 that the colour 1
refers to black). The two other attributes, `type` and `data`, should be familiar;
however, the `type` attribute in `xyplot` has more options than in the standard
`plot` function. For example, `type` = " r" adds a regression line, `type` =
" smooth " adds a LOESS fit, `type` = " g" adds a reference grid, `type` = " l"
draws a line between the points, and `type` ="a" adds a line connecting the means
of groups within a panel.

The `strip` argument should contain a logical (either TRUE or FALSE),
meaning either do, or do not, draw strips, or a function giving the necessary
input (in this case `strip.default`). To see what these options do, run the
basic `xyplot` command.

```
> xyplot(SAL ~ MyTime |factor(Station), data = Env)
```

Compare this with (results are not shown here):

```
> xyplot(SAL ~ MyTime | factor(Station), type = "l",
       strip = TRUE, col.line = 1, data = Env)
```

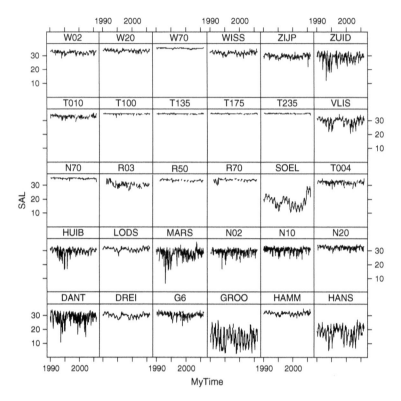

Fig. 8.1 Multipanel plot showing salinity (SAL) at 30 stations along the Dutch coast over 15 years. Note the differences among stations in spread and average values

```
> xyplot(SAL ~ MyTime | factor(Station), type = "l",
         strip = FALSE, col.line = 1, data = Env)
```

From the graph in Fig. 8.1 we note that some stations have generally lower salinity levels. Water in the North Sea has a salinity of around 32, probably because of proximity to rivers or other sources of fresh water inflow. Salinity values vary among stations, with stations that have lower values showing greater fluctuations over time. Another point to note is that some stations show similar patterns; possibly these stations are located near one another. It is difficult to see whether there is a seasonal pattern in these data. To investigate this, we can utilise the lattice function, bwplot, to draw box-and-whisker plots.

Do Exercise 2 in Section 8.11. This is an exercise in using the xyplot function with a temperature dataset.

8.3 Multipanel Boxplots: **bwplot**

A box-and-whisker plot, or boxplot, of the salinity data is shown in Fig. 8.2.
The function boxplot was introduced in Chapter 7. The multipanel counter-
part is called bwplot and uses a formula layout similar to the function xyplot.
This time, however, we plot Salinity against Month (numbered 1–12) and
our conditioning variable is not Station, but Area. There are two reasons for
doing this. First, we have seen that some stations show similar patterns and we
know that these are located in the same area. The second reason is that there are
not sufficient data per station for each month to draw meaningful box-and-
whisker plots, so we combine stations and years. Hence, the panels show the
median and spread of the salinity data for each month in each of ten areas. Here
is the code.

```
> setwd("C:/RBook")
> Env <- read.table(file ="RIKZENV.txt", header = TRUE)
> library(lattice)
> bwplot(SAL ~ factor(Month) | Area,
       strip = strip.custom(bg = 'white'),
       cex = 0.5, layout = c(2, 5),
       data = Env, xlab = "Month", ylab = "Salinity",
       par.settings = list(
          box.rectangle = list(col = 1),
          box.umbrella = list(col = 1),
          plot.symbol = list(cex = .5, col = 1)))
```

The code appears extensive but could have been shorter if we had not wanted
to draw all items in the graph in black and white (by default colours are used). If
colours and labels are not a consideration use (results are not presented here):

```
> bwplot(SAL ~ factor(Month) | Area, layout = c(2, 5),
       data = Env)
```

However, this graph is not as appealing, and we continue with the more
extensive code. The list following par.settings is used to set the colour
of the box, the whiskers (called umbrella), and the size and colour of the
open circles (representing the median). We again set the strip colour to
white. We use the layout argument to set the panel arrangement to a
rectangular grid by entering a numeric vector specifying the number of
columns and rows.

The variability in the data, as displayed in Fig. 8.2, differs among the
areas. There also appears to be a cyclic component, probably illustrating a
seasonal effect (e.g., river run-off); however, this is not equally clear for all
areas.

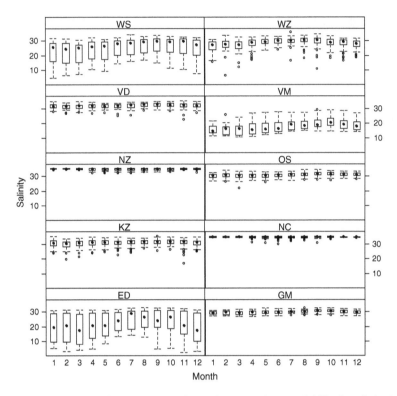

Fig. 8.2. Multipanel plot showing salinity for regions over time. Variability in salinity levels differs among regions

Do Exercise 3 in Section 8.11 in the use of the `bwplot` function using the temperature data.

8.4 Multipanel Cleveland Dotplots: `dotplot`

The Cleveland dotplot, called `dotplot` in lattice, was introduced in Chapter 7 as `dotchart`. Because there are so many data points in the salinity dataset, we restrict our plot to stations in a single area. The following code produces a multipanel dotplot, and the resulting graph is in Fig. 8.3.

```
> setwd("C:/RBook")
> Env <- read.table(file ="RIKZENV.txt", header = TRUE)
> library(lattice)
```

```
> dotplot(factor(Month) ~ SAL | Station,
    subset = Area=="OS", jitter.x = TRUE, col = 1,
    data = Env, strip = strip.custom(bg = 'white'),
    cex = 0.5, ylab = "Month", xlab = "Salinity")
```

The code is similar to that of the xyplot and bwplot functions. We reversed the order of salinity and month in the formula to ensure that salinity is plotted along the horizontal axis and month along the vertical axis (so that it matches the interpretation of the dotchart function; see Chapter 7).

There are two additional arguments in the code, subset and jitter.x. The subset option was used to create a subselection of the data. The OS stands for the area, Oosterschelde, and jitter.x = TRUE adds a small amount of random variation in the horizontal direction to show multiple observations with the same value in the same month.

Figure 8.3 shows data points that appear to be outside the normal range, potential outliers. It may be advisable to remove these before doing statistical analyses. However, this is a subjective choice that should not be made lightly.

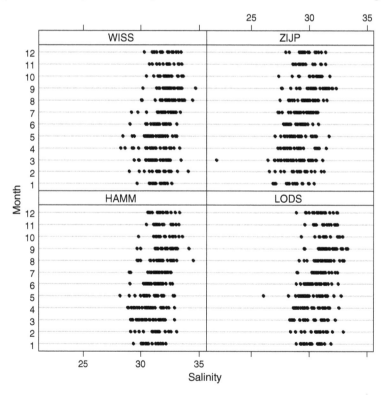

Fig. 8.3 Multipanel dotplot showing the salinity data for the four stations in the OS area. Each data point is displayed as a dot. The *y*-axis shows the month and the *x*-axis the salinity values. Note the two potential outliers in stations ZIJP and LODS

It is the responsibility of the owner of the data to make sure that data removal can be justified. It may be that the two low salinity values were the result of excessive rainfall. If the intent is to relate precipitation with salinity we might want to keep these data points.

 Do Exercise 4 in Section 8.11 in the use of the multipanel `dotplot` function using temperature data.

8.5 Multipanel Histograms: `histogram`

The function `histogram` in the lattice package can be used to draw multiple histograms. The code below draws Fig. 8.4.

```
> setwd("C:/RBook")
> Env <- read.table(file ="RIKZENV.txt", header = TRUE)
> library(lattice)
> histogram( ~ SAL | Station, data = Env,
      subset = (Area == "OS"), layout = c(1, 4),
      nint = 30, xlab = "Salinity", strip = FALSE,
      strip.left = TRUE, ylab = "Frequencies")
```

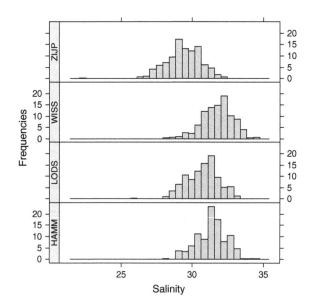

Fig. 8.4 A lattice `histogram` of salinity data for stations of the OS area

Note the slightly different format of the formula (we only need salinity data to plot the histograms). Again, we only present data for four stations using `subset`. We have changed the `layout` so that the panels are arranged vertically and increased the number of bars, the so-called bins, to 30 with the `nint` argument, as we found the default number to be too few. We also moved the strip to the side of the panels by setting `strip = FALSE` and `strip.left = TRUE`. Within the OS area there appears to be one station, ZIJP, with generally lower salinity.

To create a density plot, change the function name `histogram` to `densityplot`. If the argument for the number of bins is not removed, R will ignore it. Another function for plotting data distributions is `qqmath`, which draws QQ-plots. This stands for Quantile–Quantile plots, which are used to compare distributions of continuous data to a theoretical distribution (most often the Normal distribution).

8.6 Panel Functions

Panel functions were introduced in Chapter 7 with `pairs` and `coplot`. Remember that they are ordinary functions (see Chapter 6) that are used to draw the graph in more than one panel.

Panel functions in lattice are executed automatically within each high-level lattice function. As mentioned in Section 8.1, each default panel function contains the name of its "parent" function, for example, `panel.xyplot`, `panel.bwplot`, `panel.histogram`, and so on. Thus, when you type `xyplot(y ~ x | z)`, R executes: `xyplot (y ~ x | z, panel = panel.xyplot)`. The argument `panel` is used to associate a specific panel function with the plotting regime. Because a panel function is a function we could have written:

```
xyplot (y ~ x | z, panel = function (...) {
        panel.xyplot(...) })
```

The "..." argument is crucial, as it is used to pass on information to the other functions. Apart from `y,` `x,` and `z,` `xyplot` calculates a number of parameters before doing the actual plotting, and those that are not recognized are handed down to the `panel` function where they are used if requested. The consequence is that you can provide arguments to the panel functions at the level of the main function as well as within the panel function. You can write your own panel functions, but lattice contains a number of predefined functions that are easier to use. Panel functions can, and often do, call other panel functions, depending on the arguments. We discuss three examples of the use of panel functions.

8.6.1 First Panel Function Example

This example again uses the salinity dataset, this time to explore the potential relationship between rainfall and salinity. There are no precipitation data, so we use Month as a continuous variable, assuming that rainfall is linked to time of year. We restrict the data to a single station (GROO) and condition this subset on Year. Within xyplot we call three panel functions: panel.xyplot, panel.grid, and panel.loess. We set limits for Month of 1–12 and for Salinity of 0–30.

```
> setwd("C:/RBook")
> Env <- read.table(file ="RIKZENV.txt", header = TRUE)
> library(lattice)
> xyplot(SAL ~ Month | Year, data = Env,
    type = c("p"), subset = (Station =="GROO"),
    xlim = c(0, 12), ylim = c(0, 30), pch = 19,
    panel = function (...) {
    panel.xyplot(...)
    panel.grid(..., h = -1, v = -1)
    panel.loess(...) })
```

The resulting graph is presented in Fig. 8.5. Note how the points are on the gridlines. This is because panel.grid comes after panel.xyplot in the panel function. If you reverse the order of panel.grid and panel.xyplot, the grid is automatically drawn first. The panel function panel.loess adds a smoothing line. The amount of smoothing can be controlled by adding span = 0.9 (or any other value between 0 and 1) as a main attribute to xyplot (see Hastie and Tibshiranie (1990) for details on LOESS smoothing and span width).

We included options in the panel.grid function to align the vertical and horizontal gridlines with the axes labels. A positive number for h and v specifies the number of horizontal and vertical gridlines. If negative values for h and v are specified, R will try to align the grid with the axes labels. Experiment with different values for h and v, and see what happens.

Another important point is that, without including panel.xyplot in the code, the data points will not be plotted. Because Year is interpreted as a continuous variable, the strip has a different format than if Year were a factor. The year is represented by a coloured vertical bar in the strip. This is not very useful, and it is probably advisable to define year as a factor, so that it will print the values for year in the strips.

The data show clear signs of seasonality, although there is apparent variation in the annual salinity patterns. Nearly the same figure can be obtained with:

```
> xyplot(SAL ~ Month | Year, data = Env,
    subset = (Station == "GROO"), pch = 19,
    xlim = c(0, 12), ylim = c(0, 30),
    type = c("p", "g", "smooth"))
```

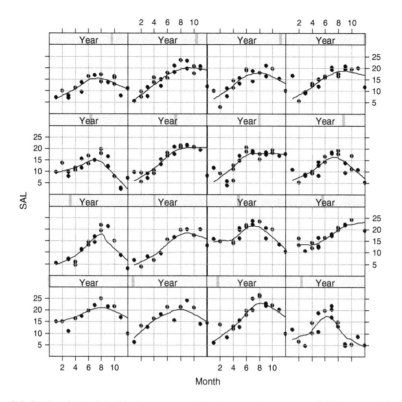

Fig. 8.5 Scatterplots of `Salinity` versus `Month` over the course of 16 years, with the addition of a *grid* and a *smoothing line*. The data show a clear seasonal pattern. Because Year is not defined as a factor, they are represented by *vertical lines* in the strips

Note that the `type` argument has the values `"p"` , `"g"` and `"smooth"`. As a result, the `xyplot` function executes the `panel.xyplot`, `panel.grid`, and `panel.smooth` functions.

8.6.2 Second Panel Function Example

The second example presents the multipanel Cleveland dotplot shown in Fig. 8.3, this time using a different colour and increased size for the dots representing potential outliers. The graph is shown in Fig. 8.6. Because this book is in greyscale, the two larger red points are printed in black.

Figure 8.6 can be created by two methods. The first option is to use the same code as in Section 8.4, and add the code `cex = MyCex` as the main argument, where `MyCex` is a vector of the same length as SAL with predefined values for `cex`. The second option is to determine values for `cex` in the panel function. The following demonstrates the second approach.

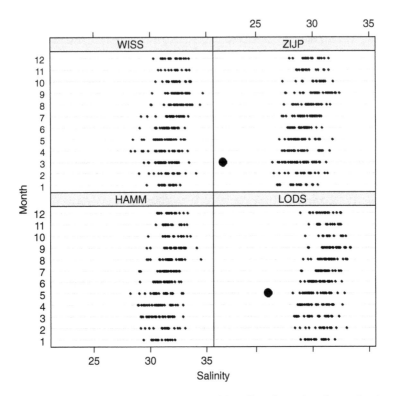

Fig. 8.6 Multipanel Cleveland dotplot with potential outliers shown by a larger dot size

A cut-off level for increasing the point size and changing its colour was set at salinity lower than the median minus three times the difference between the third and first quartiles. Note that this is a subjective cut-off level. The following code was used.

```
> setwd("C:/RBook")
> Env <- read.table(file ="RIKZENV.txt", header = TRUE)
> library(lattice)
> dotplot(factor(Month) ~ SAL | Station, pch = 16,
    subset = (Area=="OS"), data = Env,
    ylab = "Month", xlab = "Salinity",
    panel = function(x, y, ...) {
      Q <- quantile(x, c(0.25, 0.5, 0.75) ,
                    na.rm = TRUE)
      R <- Q[3] - Q[1]
      L <- Q[2] - 3 * (Q[3] - Q[1])
      MyCex <- rep(0.4, length(y))
      MyCol <- rep(1, length(y))
```

```
MyCex [x < L] <- 1.5
MyCol [x < L] <- 2
panel.dotplot(x, y, cex = MyCex,
             col = MyCol, ...)})
```

The main arguments are the formula, data, xlab, and ylab. The panel function has as arguments x, y, and "...". This means that inside the panel function, x contains the salinity data for a specific station, and y the corresponding months. Inside a panel, the x and the y constitute a subset of the data corresponding to a particular station. The "..." is used to pass on general settings such as the pch value. The quantile function is used to determine the first and third quantiles and the median. The cut-off level is specified (L), and all x values (salinity) smaller than L are plotted with cex = 1.5 and col = 2. All other values have the values cex = 0.4 and col = 1. The code can be further modified to identify considerably larger salinity values. In this case, L and x < L must be changed. We leave this as an exercise to the reader.

8.6.3 Third Panel Function Example*

This section discusses graphing tools that can be used to illustrate the outcome of a principal component analysis (PCA). It is marked with an asterisk, as the material is slightly more difficult, not with respect to the R code, but due to the use of multivariate statistics. It is an exception in being one of the few parts of this book that requires knowledge of statistics to follow the text. If the graph in Fig. 8.7 looks interesting, read on.

Figure 8.7 shows four biplots.[1] The data used here are morphometric measurements taken on approximately 1000 sparrows (Chris Elphick, University of Connecticut, USA). Zuur et al. (2007) used these data to explain PCA in detail.

The interpretation of a PCA biplot depends on various choices, and a full discussion is outside the scope of this text. See Jolliffe (2002) or Zuur et al. (2007) for details. In this case, the morphometric variables are represented as lines from the origin to a point, with coordinates given by the loadings of the first two axes. The specimens are presented as points with coordinates given by the scores of the first two axes. Depending on the chosen scaling, loading and/or scores need to be multiplied by corresponding eigenvalues (Jolliffe, 2002).

The biplot allows us to make statements of which variables are correlated, which specimens are similar, and whether specimens show high (or low) values

---

[1] A biplot is a tool to visualise the results of multivariate techniques such as principal component analysis, correspondence analysis, redundancy analysis, or canonical correspondence analysis. Using specific rules in the PCA biplot, correlations (or covariances) among the original variables, relationships among observations, and relationships between observations and variables can be inferred. There are various ways to scale the biplot, and the interpretation of the biplot depends on this scaling.

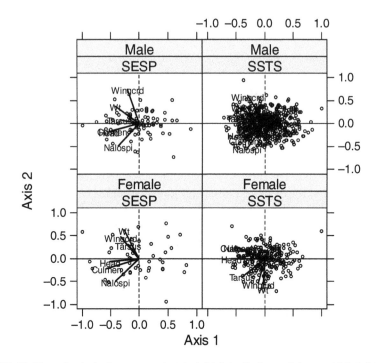

Fig. 8.7 Multipanel principal component analysis biplots. Each panel shows a biplot obtained by applying PCA (using the correlation matrix) to a dataset. SESP and SSTS represent the species seaside sparrows (*Ammodramus maritimus*) and saltmarsh sharp-tailed sparrows (*Ammodramus caudacutus*). The *graphs* indicate that the nalospi, culmen, and head measurements are correlated, and this makes sense as these are all nested subsets of each other. Wing length, mass, and tarsus on the other hand are indicators of the overall structure size of the bird, so again it makes sense that these are correlated (as suggested by the biplots), but not necessarily correlated with the first three

for particular variables. These statements are based on the directions of the lines and positions of the points. Lines pointing in a similar direction correspond to positively correlated variables, lines with an angle of 90 degrees correspond to variables that have a small correlation, and lines pointing in (approximately) opposite directions correspond to negatively correlated variables. There are also criteria for comparing points and comparing the points to the lines. The interested reader is referred to the aforementioned literature.

The sampled sparrows can be separated into two sexes and two species (SESP and SSTS). The following code was used to create Fig. 8.7.

```
> setwd("C:/RBook")
> Sparrows <- read.table(file = "Sparrows.txt",
                         header = TRUE)
```

```
> library(lattice)
> xyplot(Wingcrd ~ Tarsus | Species * Sex,
      xlab = "Axis 1", ylab = "Axis 2", data = Sparrows,
      xlim = c(-1.1, 1.1), ylim = c(-1.1, 1.1),
      panel = function(subscripts, ...){
         zi <- Sparrows[subscripts, 3:8]
         di <- princomp(zi, cor = TRUE)
         Load <- di$loadings[, 1:2]
         Scor <- di$scores[, 1:2]
         panel.abline(a = 0, b = 0, lty = 2, col = 1)
         panel.abline(h = 0, v = 0, lty = 2, col = 1)
         for (i in 1:6){
             llines(c(0, Load[i, 1]), c(0, Load[i, 2]),
                    col = 1, lwd = 2)
             ltext(Load[i, 1], Load[i, 2],
                    rownames(Load)[i], cex = 0.7)}
         sc.max <- max(abs(Scor))
         Scor <- Scor / sc.max
         panel.points(Scor[, 1], Scor[, 2], pch = 1,
                    cex = 0.5, col = 1)
      })
```

The xlab, ylab, and data arguments are familiar. The first part of the equation, Wingcrd ~ Tarsus, is used to set up the graph. There was no specific reason for choosing to use these two variables in the formula. The portion of the code following the | symbol is new. So far, we have only used one conditioning variable, but in this case there are two, Species and Sex. As a result, the lower two panels in the graph show the female data, and the upper two panels show the data from males. Change the order of Species and Sex to see what happens. Note that both variables are defined as characters in the data file; hence R automatically treats them as factors.

The xlim and ylim values require some statistical explanation. The outcome of a PCA can be scaled so that its numerical information (scores) for a graph fits between −1 and 1. See Legendre and Legendre (1998) for mathematical details.

It is also important when constructing the graph to ensure that the distance in the vertical direction is the same as in the horizontal direction to avoid distortion of the angles between lines.

We now address the more difficult aspect, the panel function. The vector subscripts automatically contains the row number of the selected data in the panel function. This allows us to manually extract the data that are being used for a certain panel using Sparrows[subscripts, 3:8]. The 3:8 designates the variables Wingcrd, Tarsus, Head, Culmen, Nalospi,

and Wt.[2] The princomp function applies principal component analysis, and
loadings and scores of the first two axes are extracted. The two panel.ab-
line functions draw the axes through the origin. The loop (see Chapter 6)
is used to draw the lines and to add the labels for all variables. The
functions llines and ltext do the work. Finally, we rescale all the scores
between −1 and 1, and add them as points with the panel.xyplot
function.

The resulting biplots show how the correlations among the morphometric
variables differ according to sex and species.

The code can easily be extended to allow for triplots obtained by redundancy
analysis or canonical correspondence analysis (see the functions rda and cca
in the package vegan).

Full details on PCA and biplots and triplots can be found in Jolliffe (2002),
Legendre and Legendre (1998), and Zuur et al. (2007), among many other sources.

 Do Exercise 6 in Section 8.11 on the use of panel functions using the
temperature data.

8.7 3-D Scatterplots and Surface and Contour Plots

Plots for displays of three variables, sometimes called trivariate displays, can be
generated by the functions cloud, levelplot, contourplot, and wire
frame. In our opinion, three-dimensional scatterplots are not always useful.
But they look impressive, and we briefly discuss their creation. Below is an
example of the function cloud, applied to the Dutch environmental dataset,
which produces three-dimensional scatterplots showing the relationships
among chlorophyll-a, salinity, and temperature. The code is fairly simple, and
the resulting graph is presented in Fig. 8.8.

```
> setwd("C:/RBook")
> Env <- read.table(file ="RIKZENV.txt", header = TRUE)
> library(lattice)
> cloud(CHLFa ~ T * SAL | Station, data = Env,
    screen = list(z = 105, x = -70),
    ylab = "Sal.", xlab = "T", zlab = "Chl. a",
    ylim = c(26, 33), subset = (Area=="OS"),
    scales = list(arrows = FALSE))
```

---

[2] In Chapter 2 we provided an explanation for Wingcrd, Tarsus, Head, and Wt. Culmen
measures the length of the top of the bill from the tip to where feathering starts, and Nalospi
the distance from the bill top to the nostril.

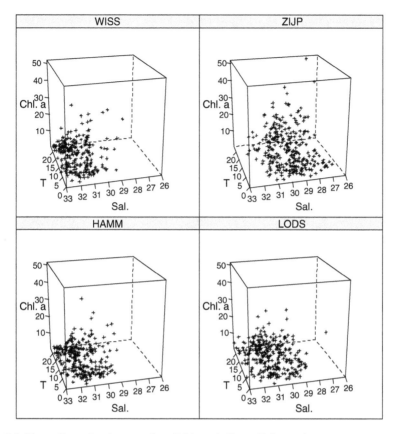

Fig. 8.8 Three-dimensional scatterplot of chlorophyll-a, salinity, and temperature

The function `cloud` uses several arguments that we have not previously introduced. The option `screen` is used to denote the amount of rotation about the axes in degrees. With `arrows = FALSE` we removed arrows that are normally plotted along the axes of three-dimensional graphs to indicate the direction in which values increase. Consequently, axis tick marks, which by default are absent, are now shown. We limited the *y*-axis values to between 26 and 33.

The functions `levelplot`, `contourplot`, and `wireframe` are used to plot surfaces. This generally involves predicting values on a regular grid by statistical functions, which is outside the scope of this book. More information on these functions can be found in their help pages.

8.8 Frequently Asked Questions

There are a number of things that we have often found ourselves modifying when making lattice plots. The following are some that we have found useful.

8.8.1 How to Change the Panel Order?

By default panels are drawn starting from the lower-left corner, proceeding to the
right, and then up. This sequence can be changed by setting as.table = TRUE
in a high-level lattice call, resulting in panels being drawn from the upper-left
corner, going right, and then down.

The order of the panels can also be changed by defining the condition
variable as a factor, and changing the level option in the factor function.
Figure 8.9 shows a multipanel scatterplot of abundance of three bird species on
three islands in Hawaii. The data were analysed in Reed et al. (2007). The
problem with the graph is that the time series are arranged randomly with
respect to species and island, which makes comparisons among time series of
bird abundance of an island, or of an individual species, more difficult.

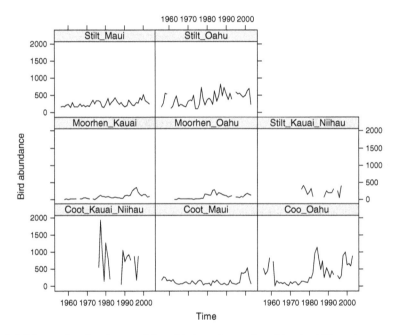

Fig. 8.9 Time series abundances of three bird species on three islands of Hawaii

Figure 8.10 on the other hand, shows the time series of each island in rows, and
time series of each species in the columns. This makes the comparison of trends
with respect to individual species or islands much easier. So, how did we do it?

The following code imports the data and uses the as.matrix and as.vec-
tor commands to concatenate the eight abundance time series into a single
long vector. The as.matrix command converts the data frame into a matrix,
which allows as.vector to make the conversion to a long vector; as.vector
will not work with a data frame. The rep function is used to create a single long
vector containing eight repetitions of the variable Year.

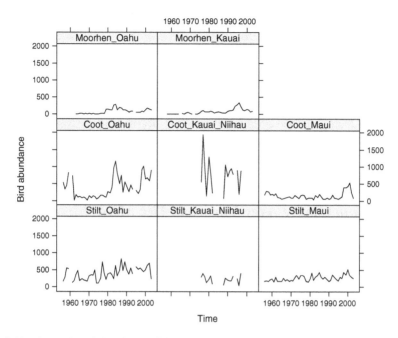

Fig. 8.10 Time series of abundance of three bird species on three islands of Hawaii. Note that time series of an island are arranged *vertically*, and time series of a species are *horizontal*

```
> setwd("C:/RBook")
> Hawaii <- read.table("waterbirdislandseries.txt",
                     header = TRUE)
> library(lattice)
> Birds <- as.vector(as.matrix(Hawaii[, 2:9]))
> Time <- rep(Hawaii$Year, 8)
> MyNames <- c("Stilt_Oahu", "Stilt_Maui",
            "Stilt_Kauai_Niihau","Coot_Oahu",
            "Coot_Maui", "Coot_Kauai_Niihau",
            "Moorhen_Oahu","Moorhen_Kauai")
> ID <- rep(MyNames, each = 48)
```

The rep function is also used to define a single long vector ID in which each name is repeated 48 times, as each time series is of length 48 years (see Chapter 2). Figure 8.9 was made with the familiar code:

```
> xyplot(Birds ~ Time | ID, ylab = "Bird abundance",
        layout = c(3, 3), type = "l", col = 1)
```

The layout option tells R to put the panels in 3 rows and 3 columns with points connected by a black line.

To change the order of the panels, change the order of the levels of the factor ID:

```
> ID2 <- factor(ID, levels = c("Stilt_Oahu",
      "Stilt_Kauai_Niihau", "Stilt_Maui",
      "Coot_Oahu", "Coot_Kauai_Niihau", "Coot_Maui",
      "Moorhen_Oahu", "Moorhen_Kauai"))
```

Note the change in the order of the names. Rerunning the same xyplot command, but with ID replaced by ID2, produces Fig. 8.10. Determining the order of the levels within the factor ID2 (the names of the bird/island combinations) is a matter of trial and error.

8.8.2 How to Change Axes Limits and Tick Marks?

The most direct way to influence the range of values on the axes is by using xlim and ylim; however, this will result in the same limits on both the x and y axes of all panels. The scales option is more versatile. It can be used to define the number of tick marks, the position and labels of ticks, and also the scale of individual panels.

In Figure 8.10 the vertical ranges of the time series differ among the panels. This is obviously because some species are more abundant than others. However, if we want to compare trends over time, we are less interested in the absolute values. One option is to standardise each time series. Alternatively, we can allow each panel to set its own range limits on the y-axis. This is done as follows (after entering the code from the previous subsection).

```
> xyplot(Birds ~ Time|ID2, ylab = "Bird abundance",
      layout = c(3, 3), type = "l", col = 1,
      scales = list(x = list(relation = "same"),
                    y = list(relation = "free")))
```

The option scales can contain a list that determines attributes of both axes. In this case, it specifies that the x-axes of all panels have the same range, but sets a vertical range of each panel appropriate to the data. The resulting graph is presented in Fig. 8.11.

To change the direction of the tick marks inwards use the following code.

```
> xyplot(Birds ~ Time|ID2, ylab = "Bird abundance",
      layout = c(3, 3), type = "l", col = 1,
      scales = list(x = list(relation = "same"),
                    y = list(relation = "free"),
                    tck = -1))
```

The tck = -1 is within the list argument of the scales option. There are many more arguments for scales; see the xyplot help file.

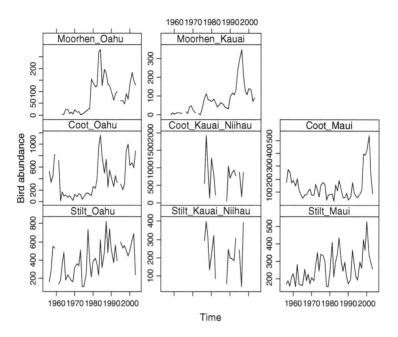

Fig. 8.11 Time series of abundance of three bird species on three islands of Hawaii. Each panel has an appropriate abundance value range

8.8.3 Multiple Graph Lines in a Single Panel

The attribute `groups` in high-level lattice functions can be used when there is a grouping in the data that is present in each level of the conditioning variable. Figure 8.12 shows all time series of a species in a single panel. The following code was used to generate the graph.

```
> Species <- rep(c("Stilt", "Stilt", "Stilt",
                    "Coot", "Coot", "Coot",
                    "Moorhen", "Moorhen"), each = 48)
> xyplot(Birds ~ Time | Species,
         ylab = "Bird abundance",
         layout = c(2, 2), type = "l", col = 1,
         scales = list(x = list(relation = "same"),
                       y = list(relation = "free")),
         groups = ID, lwd = c(1, 2, 3))
```

The first command defines a vector `Species` identifying which observations are from which species. The `xyplot` with the `groups` option then draws the time series of each species in a single panel. The option `lwd` was used to draw lines of different thickness to represent the three islands.

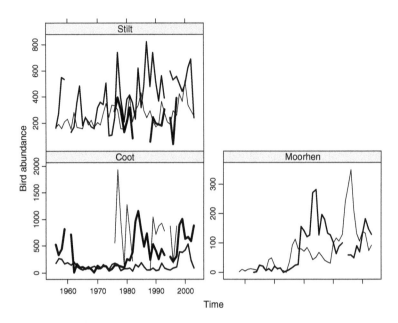

Fig. 8.12 Abundance of three species of Hawaiian birds over time. Data on each species are plotted in a single panel

Do Exercise 7 in Section 8.11 on creating multiple lines in an `xyplot` function using the temperature data.

8.8.4 Plotting from Within a Loop*

If you did not read Chapter 6, you may want to skip this section. Recall from Section 8.2, that the salinity dataset consists of time series at stations, and that the stations are located in areas along the Dutch coast. In Section 8.6.2, we created a dotplot of the data from the area OS (Figure 8.6). Suppose we want to make the same graph for each of the 12 areas. One option is to enter the code from Section 8.6.2 12 times and each time to change the `subset` option. However, in Chapter 6, we demonstrated executing similar plotting commands automatically within a loop. The only difference is that we need to replace the `plot` command by a `dotplot` command:

```
> setwd("C:/RBook")
> Env <- read.table(file ="RIKZENV.txt", header = TRUE)
> library(lattice)
> AllAreas <- levels(unique(Env$Area))
> for (i in AllAreas ){
```

```
Env.i <- Env[Env$Area == i,]
win.graph( )
dotplot(factor(Month)~SAL | Station, data = Env.i)
}
```

The first three lines load the data and the lattice package. The variable
`AllAreas` contains the names of the 12 areas. The loop iteratively extracts
the data of an area and draws a dotplot of all stations in this area. The only
problem is that this code will produce 12 empty graph windows.

When you execute a high-level lattice function, a graph is created on your
screen. This appears similar to using, for example, the traditional `plot` command. The lattice command differs, however, because this command returns an
object of class "trellis," and, in order to see a plot, the `print` function is invoked.
Sometimes, when issuing a command to draw a lattice plot, nothing happens, not
even an error message. This most often happens when creating a lattice plot inside
a loop or function, or from a `source` command. To get the graphs, the `print`
command must be embedded in the loop:

```
print(dotplot(factor(Month)~SAL | Station,
      data = Env.i))
```

Adding the `print` command to the code and rerunning it will produce 12
windows with graphs.

8.8.5 Updating a Plot

Because drawing lattice plots is time consuming, especially when you are new to
lattice, the `update` function is useful. Many attributes of a lattice object can be
changed with `update`, thus your graph must be stored in an object first. An
additional advantage is that if you experiment by using the `update` command,
your original graph is not changed, so

```
> MyPlot <- xyplot(SAL ~ MyTime | Station,
                type = "l", data = Env)
> print(MyPlot)
> update(MyPlot, layout = c(10, 3))
```

will print the plot in a new layout. The `update` command will automatically
generate a plot because it is not assigned to an object. The original object
`MyPlot` is unchanged.

8.9 Where to Go from Here?

After completing the exercises you will have the flavour of lattice graphs and
will undoubtedly want to use them in research, publications, and presentations. For further information consult Sarkar (2008) or Murrell (2006). Other

sources are the website that accompanies Sarkar (2008) (http://lmdvr.r-forge.
r-project.org) or the R-help mailing list.

8.10 Which R Functions Did We Learn?

Table 8.2 contains the R functions introduced in this chapter.

Table 8.2 R functions introduced in this chapter

Function	Purpose	Example
xyplot	Draws a scatterplot	xyplot (y ~ x | g, data = data)
histogram	Histogram	histogram(~ x | g, data = data)
bwplot	Comparative box-and-whisker plots	bwplot(y ~ x | g, data = data)
dotplot	Cleveland dotplot	dotplot(y ~ x | g, data = data)
cloud	Three-dimensional scatterplot	cloud(z ~ x * y | g, data = data)

8.11 Exercises

Exercise 1. Using the `demo(lattice)` function.
    Load the lattice package and investigate some of the possibilities by typing in
`demo(lattice)`. Type in `?xyplot` and copy and paste some of the examples.

Exercise 2. Using the `xyplot` with temperature data.
    Create a multipanel scatterplot in which temperature is plotted versus time
for each station. What is immediately obvious? Do the same for each area.
What goes wrong and how can you solve this? Add a smoother and a grid to
each panel.

Exercise 3. Using the `bwplot` with temperature data.
    Create a boxplot in which temperature is plotted versus month for each area.
Compare with the boxplot for the salinity data and comment on the differences
in the patterns.

Exercise 4. Using the `dotplot` function with salinity data.
    Use Cleveland dotplots to discover if there are more outliers in the salinity
data, making a lattice plot with all stations as panels. Compare with Fig. 8.3.
What can be noted on the scale of the y-axis? Look up the argument `relation`
in the help page of `xyplot` and use it.

Exercise 5. Using the density plot with salinity data.
    Change Fig. 8.4 to a density plot. Is it an improvement? Add the following
argument: `plot.points = "rug"`. To compare density distributions you
might prefer to have all the lines in a single graph. This is accomplished with the

groups argument. Remove the conditioning argument and add groups = Station (see also Section 8.8). Add a legend to specify the lines representing each station. This requires advanced programming (though there are simple solutions), and we refer you to the code on our website for the solution.

Exercise 6. Using the xyplot function with temperature data.

Look at the help pages of panel.linejoin. Create a plot similar to Fig. 8.2, but with temperature on the *y*-axis. This is the same as in Exercise 3, but now use panel.linejoin to connect the medians, not the means. Take care of the NAs in the data, otherwise nothing will happen.

Exercise 7. Using the xyplot function with salinity data.

Create a lattice scatterplot using salinity as the dependent variable versus time for each area and include the groups argument to draw separate lines for each station.

Exercise 8. Using the xyplot function with temperature data.

In Exercise 2 you created a lattice scatterplot for each area using temperature as the dependent variable versus time. Make a similar graph for the area "KZ", but plot small dots and add a smoothing line with 1/10 span width. Create strips on either side of the panels, with the text "Area 1", "Area 2", and so on. Add an *x*-label, a *y*-label, and a title.

Exercise 9. Using the xyplot function with salinity data.

Create a multipanel scatterplot of the salinity data versus time conditional on area with different lines (no points) for the different stations within each area. Make sure the panel layout is in two columns. Use the same *x*-axis on each panel, but different scales for the *y*-axes. Limit the number of tick marks and labels on the *y*-axes to three or four and on the *x*-axes to four, with the tick marks between labels. Remove the tick marks (and labels) from the top and make sure they are only present on the bottom of the graph. Decrease the size of the text in the strip and the height of the strip. Add a grid (properly aligned with the tick marks), and also *x*- and *y*-labels. Change the order of the panels to alphabetic from top left to bottom right.

Exercise 10. Using the xyplot function with the ISIT data.

Create a multipanel scatterplot of the ISIST data (see Chapter 1). Plot the sources versus depth for each station. Also make a multipanel graph in which data from all stations sampled in the same season are grouped (see also Exercise 4 in Section 3.7). Each panel (representing a season) should have multiple lines.

Chapter 9
Common R Mistakes

The following addresses avoiding some errors that we see on a regular basis during our R courses.

9.1 Problems Importing Data

9.1.1 Errors in the Source File

The code required to import data into R was discussed in Chapter 2. The first major task is ensuring that the spreadsheet (or ascii file) is adequately prepared. Do not use spaces in variable names or include blank cells. The error messages that will result were shown in Chapter 2, and are not repeated here.

If your column names are species names of the form *Delphinus delphi*, call it *Delphinus.delphi* with a point between the two names, *Delphinus_delphi* (underscore), or, better yet, something shorter, such as *Ddelphi*.

9.1.2 Decimal Point or Comma Separation

Another potential pitfall is the decimal separation used: comma or point. We often teach groups in which some participants have computers with point separation and the rest use commas. In Chapter 2, we demonstrated use of the dec option in the read.table function to set the style of separation. *Always* use the str function after importing the data to verify that the data have been imported as intended. If you import the data using the incorrect dec option, R will accept it without an error message. The difficulties arise later when you attempt to work with the data, for example, to make a boxplot or take the mean of a variable which is continuous but has erroneously been imported as a categorical variable because of the wrong dec option.

The problems may be compounded by the fact that the mistake is not always readily apparent, because you may occasionally get away with using the wrong decimal separator. In the following example, the first two commands import the

A.F. Zuur et al., *A Beginner's Guide to R*, Use R,
DOI 10.1007/978-0-387-93837-0_9, © Springer Science+Business Media, LLC 2009

cod parasite data that were used in Chapter 6. Note that we used the dec = ","
option in the read.table command, although the ascii file contains data with
decimal point separation.

```
> setwd("c:/RBook/")
> Parasite <- read.table(file = "CodParasite.txt",
                 header = TRUE, dec = ",")
```

The str function shows the imported data:

```
> str(Parasite)
'data.frame'  : 1254 obs. of 11 variables:
$Sample     : int 1 2 3 4 5 6 7 8 9 10 ...
$Intensity  : int 0 0 0 0 0 0 0 0 0 0 ...
$Prevalence : int 0 0 0 0 0 0 0 0 0 0 ...
$Year       : int 1999 1999 1999 1999 1999 ...
$Depth      : int 220 220 220 220 220 220 220 ...
$Weight     : Factor w/ 912 levels "100",..: 159...
$Length     : int 26 26 27 26 17 20 19 77 67 ...
$Sex        : int 0 0 0 0 0 0 0 0 0 0 ...
$Stage      : int 0 0 0 0 0 0 0 0 0 0 ...
$Age        : int 0 0 0 0 0 0 0 0 0 0 ...
```

Length has been correctly imported as an integer, but the variable Weight
is considered a categorical variable. This is because some of the weight
values are written with decimals (e.g., 148.0), whereas all other variables
are coded as integers in the text file. This means that the following
commands will work.

```
> mean(Parasite$Intensity, na.rm = TRUE)
[1] 6.182957
> boxplot(Parasite$Intensity) #Result not shown here
```

However, entering the same code for Weight gives error messages:

```
> mean(Parasite$Weight)
[1] NA
Warning message:
In mean.default(Parasite$Weight): argument is not numeric
or logical: returning NA

> boxplot(Parasite$Weight)

Error in oldClass(stats) <- cl: adding class "factor" to
an invalid object
```

If you use `Weight` as a covariate in linear regression, you may be surprised at the large number of regression parameters that it consumes; `Weight` was automatically fitted as a categorical variable. It is only by chance that the mean and boxplot functions using `Intensity` were accurately produced; if it had contained values including decimals, the same error message would have appeared.

9.1.3 Directory Names

Problems may also arise when importing data with directory names containing non-English alphabetical characters such as á, , , and many more. This is a language issue that may be difficult to resolve if you are working with datasets contributed by colleagues using other alphabet systems. Surprisingly, problems do not occur on all computers. It is advisable to keep the directory structure simple and avoid characters in the file and directory names that your computer may see as "strange".

9.2 Attach Misery

When conducting a course, we face the dilemma of whether to teach a quick-and-easy approach to accessing variables in a data frame using the `attach` function, to put participants through some extra R coding using the `data` argument (when applicable), or to teach the use of the `$` notation. This is a slightly controversial area, as some authorities insist that the `attach` function should absolutely never be used, whereas others have written books in which the function is used extensively (e.g., Wood, 2006). When we work with a single dataset, we use the `attach` command, as it is more convenient. However, there are rules that must be followed, and we see many R novices breaking these rules.

9.2.1 Entering the Same attach Command Twice

The most common problem incurred with the `attach` function arises when one program's code containing the `attach` command runs it in R, detects a programming mistake, fixes it, and proceeds to rerun the entire piece of code. Here is an example:

```
> setwd("c:/RBook/")
> Parasite <- read.table(file = "CodParasite.txt",
                  header = TRUE)
> attach(Parasite)
> Prrrarassite
Error: object "Prrrarassite" not found
```

Because we misspelled *Parasite*, R gives an error message. The obvious response is to correct the typing error and try again. However, if we correct the mistake in the text editor (e.g., Tinn-R) and then resend, or copy, all the code (which includes the `attach` command) to R, it will result in the following.

```
> setwd("c:/RBook/")
> Parasite <- read.table(file = "CodParasite.txt",
                    header = TRUE)
> attach(Parasite)
The following object(s) are masked from Parasite (posi-
tion 3):
Age Area Depth Intensity Length Prevalence Sample Sex
Stage Weight Year
```

At this point it is useful to consult the help file of the `attach` function. It says that the function adds the data frame `Parasite` to its search path, and, as a result, the variables in the data frame can be accessed without using the $ notation. However, by attaching the data frame twice, we have made available two copies of each variable. If we make changes to, for example, `Length` and, subsequently, use `Length` in a linear regression analysis, we have no way of ensuring that the correct value is used.

The alternative is to use the `detach` function before rerunning `attach` (the code below assumes that we have not yet used the attach function):

```
> setwd("c:/RBook/")
> Parasite <- read.table(file = "CodParasite.txt",
                    header = TRUE)
> attach(Parasite)
```

We can now begin programming; to detach the data frame `Parasite`, use:

```
> detach(Parasite)
```

Another procedure to avoid is running a `for` loop that in each iteration attaches a data frame, as this will generate an expanding search path which will eventually slow down your computer. The help file for the `with` function also provides an alternative to the `attach` function.

9.2.2 Attaching Two Data Frames Containing the Same Variable Names

Suppose we import the cod parasite data and the squid data and employ the `attach` function to make variables in both data frames available, using the following code.

```
> setwd("c:/RBook/")
> Parasite <- read.table(file = "CodParasite.txt",
                  header = TRUE)
> Squid <- read.table(file = "Squid.txt", header=TRUE)
> names(Parasite)

[1] "Sample"   "Intensity"   "Prevalence"   "Year"
[5] "Depth"    "Weight"      "Length"       "Sex"
[9] "Stage"    "Age"         "Area"

> names(Squid)

[1] "Sample" "Year" "Month" "Location" "Sex"
[6] "GSI"

> attach(Parasite)
> attach(Squid)

The following object(s) are masked from Parasite:
Sample Sex Year

> boxplot(Intensity ~ Sex)

Error in model.frame.default(formula=Intensity ~ Sex):
variable lengths differ (found for 'Sex')

> lm(Intensity ~ Sex)

Error in model.frame.default(formula = Intensity ~ Sex,
drop.unused.levels = TRUE): variable lengths differ
(found for 'Sex')
```

The first three commands import the data. The output of the two names functions show that both data frames contain the variable Sex. We used the attach function to make the variables in both data frames available. To see the effect of this, we can make a boxplot of the Intensity data conditional on Sex. The error message generated by the boxplot function shows that the length of the vectors Intensity and Sex differ. This is because R has used Intensity from the Parasite data frame and Sex from the Squid data frame. Imagine what would have happened if, fortuitously, these two variables were of the same dimension: we would have modelled the number of parasites in cod measured in the Barents Sea as an effect of sex of squid from the North Sea!

9.2.3 Attaching a Data Frame and Demo Data

Many statistics textbooks come with a package that contains datasets used in the book, for example, the MASS package from Venables and Ripley (2002), the

nlme package from Pinheiro and Bates (2000), and the mgcv package from
Wood (2006), among many others. The typical use of such a package is that the
reader accesses the help file of certain functions, goes to the examples at the end
of a help file, and copies and runs the code to see what it does. Most often, the
code from a help file loads a dataset from the package using the data function,
or it creates variables using a random number generator. We have also seen help
file examples that contain the attach and detach functions. When using
these it is not uncommon to neglect to copy the entire code; the detach
command may be omitted, leaving the attach function active. Once you
understand the use of the demonstrated function, it is time to try it with your
own data. If you have also applied the attach function to your data, you may
end up in the scenario outlined in the previous section.

Problems may also occur if demonstration data loaded with the data function contain some of the same variable names used as your own data files.

The general message is to be careful with the attach function, and use clear
and unique variable names.

9.2.4 Making Changes to a Data Frame After Applying
the attach Function

The following example is something that we see on a regular basis. Often, our
course participants own multiple R books, which may recommend different R
styles. For example, one book may use the attach function, whereas another
uses a more sophisticated method of accessing variables from a data frame.
Mixing programming styles can sometimes cause trouble, as can be seen from
the example below.

```
> setwd("c:/RBook/")
> Parasite <- read.table(file = "CodParasite.txt",
                 header = TRUE)
> Parasite$fSex <- factor(Parasite$Sex)
> Parasite$fSex
 [1] 0 0 0 0 0 0 0 0 0 0 0 0 0 0 0 0 0 0 0 0
[21] 0 0 0 0 0 0 0 0 0 0 0 0 0 0 0 0 0 2 1 1 1
  ...
> attach(Parasite)
> fSex
 [1] 0 0 0 0 0 0 0 0 0 0 0 0 0 0 0 0 0 0 0 0
[21] 0 0 0 0 0 0 0 0 0 0 0 0 0 0 0 0 0 2 1 1 1
  ...
> Parasite$fArea <- factor(Parasite$Area)
> fArea
Error: object "fArea" not found
```

On the first three lines, the data are imported and a new categorical variable fSex is created inside the data frame Parasite. We then make all variables in this data frame available with the attach function, and, consequently, we can access the fSex variable by simply typing it in. The numerical output shows that this was successful. If we subsequently decide to convert Area into a new categorical variable, fArea inside the data frame Parasite, we encounter a problem. We cannot access this variable by typing its name into the console (see the error message). This is because the attach function has been executed, and variables added to Parasite afterwards are not available. Possible solutions are:

1. Detach the data frame Parasite, add fArea to the data frame Parasite, and attach it again.
2. Define fArea before using the attach function.
3. Define fArea outside the data frame.

9.3 Non-attach Misery

In addition to the attach function, there are various other options available for accessing the variables in a data frame. We discussed the use of the data argument and the $ symbol in Chapter 2. In the latter case, we can use

```
> setwd("c:/RBook/")
> Parasite <- read.table(file = "CodParasite.txt",
                header = TRUE)
> M0 <- lm(Parasite$Intensity ~
            Parasite$Length * factor(Parasite$Sex))
```

The first two lines import the cod parasite data. The last two lines apply a linear regression model in which Intensity is modelled as a function of length and sex of the host. We do not discuss linear regression nor its output here. It is sufficient to know that the function has the desired effect; type summary(M0) to see the output. Note that we used the Parasite $ notation to access variables in the data frame Parasite (see Chapter 2). The following two commands load the nlme package and apply linear regression using the generalised least squares function gls (Pinheiro and Bates, 2002).

```
> library(nlme)
> M1 <- gls(Parasite$Intensity ~
            Parasite$Length * factor(Parasite$Sex))
Error in eval(expr, envir, enclos): object "Intensity"
not found
```

The results obtained by the lm and gls functions should be identical, yet R (regardless of the version used) gives an error message for the latter.

The solution is to use the data argument and avoid using the Parasite$ notation in the gls function.

9.4 The Log of Zero

The following code looks correct. It imports the dataset of the cod parasite data and applies a logarithmic transformation on the number of parasites in the variable Intensity.

```
> setwd("c:/RBook/")
> Parasite <- read.table(file = "CodParasite.txt",
                header = TRUE)
> Parasite$LIntensity <- log(Parasite$Intensity)
```

There is no error message, but, if we make a boxplot of the log-transformed values, problems become apparent; see the left boxplot in Fig. 9.1. The difficulty arises because some fish have zero parasites, and the log of zero is not defined, as can be seen from inspecting the values:

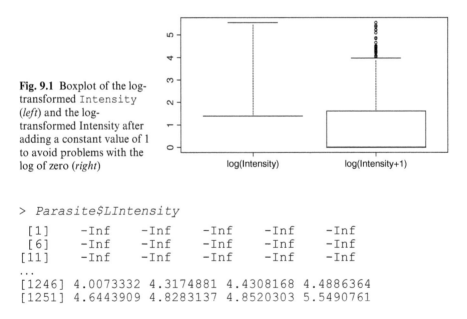

Fig. 9.1 Boxplot of the log-transformed Intensity (*left*) and the log-transformed Intensity after adding a constant value of 1 to avoid problems with the log of zero (*right*)

```
> Parasite$LIntensity
  [1]    -Inf      -Inf      -Inf      -Inf      -Inf
  [6]    -Inf      -Inf      -Inf      -Inf      -Inf
 [11]    -Inf      -Inf      -Inf      -Inf      -Inf
...
[1246]  4.0073332  4.3174881  4.4308168  4.4886364
[1251]  4.6443909  4.8283137  4.8520303  5.5490761
```

Carrying out linear regression with the variable LIntensity results in a rather intimidating error message:

```
> M0 <- lm(LIntensity ~ Length * factor(Sex),
          data = Parasite)
```

```
Error in lm.fit(x, y, offset  =  offset, singular.ok =
singular.ok,...): NA/NaN/Inf  in foreign function call
(arg 4)
```

The solution is to add a small constant value to the `Intensity` data, for example, 1. Note that there is an on-going discussion in the statistical community concerning adding a small value. Be that as it may, you cannot use the log of zero when doing calculations in R. The following code adds the constant and draws the boxplot shown on the right side in Fig. 9.1.

```
> Parasite$L1Intensity <- log(Parasite$Intensity + 1)
> boxplot(Parasite$LIntensity, Parasite$L1Intensity,
        names = c("log(Intensity)", "log(Intensity+1)"))
```

To reiterate, you should not take the log of zero!

9.5 Miscellaneous Errors

In this section, we present some trivial errors that we see on a regular basis.

9.5.1 The Difference Between 1 and l

Look at the following code. Can you see any differences between the two plot functions? The first one is valid and produces a simple graph; the second plot function gives an error message.

```
> x <- seq(1, 10)
> plot(x, type = "l")
> plot(x, type = "1")
Error in plot.xy(xy, type, ...) : invalid plot type '1'
```

The text in the section title may help to answer the question, as its font shows more clearly the difference between the l (one) and the l ("ell"). In the first function, the l in `type = "l"` stands for line, whereas, in the second plot function, the character in `type = "1"` is the numeral 1 (this is an R syntax error). If this text is projected on a screen in a classroom setting, it is difficult to detect any differences between the l and 1.

9.5.2 The Colour of 0

Suppose you want to make a Cleveland dotplot of the variable `Depth` in the cod parasite data to see the variation in depths from which fish were sampled (Fig. 9.2A). All fish were taken from depths of 50–300 meters. In addition to the numbers of parasites, we also have a variable, Prevalence, which indicates the presence (1) or absence (0) of parasites in a fish. It is interesting to add this information to the Cleveland dotplot, for example, by using different colours to denote Prevalence. This is shown in panel B. The code we use is as follows (assuming the data to have been imported as described in previous sections).

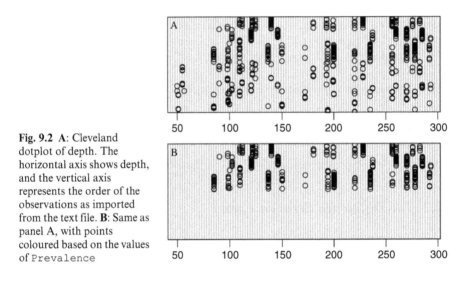

Fig. 9.2 A: Cleveland dotplot of depth. The horizontal axis shows depth, and the vertical axis represents the order of the observations as imported from the text file. **B**: Same as panel A, with points coloured based on the values of Prevalence

```
> par(mfrow = c(2, 1), mar = c(3, 3, 2, 1))
> dotchart(Parasite$Depth)
> dotchart(Parasite$Depth, col = Parasite$Prevalence)
```

We encounter a problem, in that some of the points have disappeared. This is because we used a variable in the col option that has values equal to 0, which would represent a lack of colour. It is better to use something along the lines of col = Parasite$Prevalence + 1, or define a new variable using appropriate colours.

9.5.3 Mistakenly Saved the R Workspace

Last, but not least, we deal with problems arising from mistakenly saving the workspace. Suppose that you loaded the owl data that was used in Chapter 7:

```
> setwd("C:/RBook/")
> Owls <- read.table(file = "Owls.txt", header= TRUE)
```

To see which variables are available in the workspace, type:

```
> ls()
[1] "Owls"
```

The `ls` command gives a list of all objects (after an extended work session, you may have a lot of objects).

You now decide to quit R and click on **File -> Exit**. The window in Fig. 9.3 appears. We always advise choosing "No," not saving, instead rerunning the script code from the text editor (e.g., Tinn-R) when you wish to work with it again. The only reason for saving the workspace is when running the calculations is excessively time consuming. It is easy to end up with a large number of saved workspaces, the contents of which are complete mysteries. In contrast, script code can be documented.

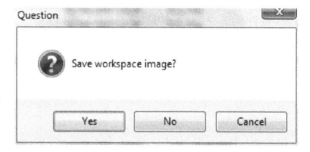

Fig. 9.3 Window asking the user whether the workspace should be saved before closing R

However, suppose that you do click on "Yes." Dealing with this is easy. The directory C:/RBook will contain a file with the extension .RData. Open Windows Explorer, browse to the working directory (in this case: C:/RBook) and delete the file with the big blue R.

Things are more problematical if, instead of using the `setwd` command, you have entered:

```
> Owls <- read.table(file = "C:/RBook/Owls.txt",
                     header = TRUE)
```

If you now quit, saving the workspace, when R is started again the following text will appear.

```
R version 2.7.2 (2008-08-25)
Copyright (C) 2008 The R Foundation for Statistical
Computing

ISBN 3-900051-07-0
R is free software and comes with ABSOLUTELY NO WARRANTY.
```

```
You are welcome to redistribute it under certain condi-
tions.

Type 'license()' or 'licence()' for distribution
details.

 Natural language support but running in an English
locale
R is a collaborative project with many contributors.
Type 'contributors()' for more information and

'citation()' on how to cite R or R packages in publica-
tions.
Type 'demo()' for some demos, 'help()' for on-line help,
or
'help.start()' for an HTML browser interface to help.

Type 'q()' to quit R.
[Previously saved workspace restored]
>
```

It is the last line that spoils the fun. R has loaded the owl data again. To convince yourself, type:

```
> Owls
```

The owl data will be displayed. It will not only be the owl data that R has saved, but also all other objects created in the previous session. Restoring a saved workspace can cause the same difficulties as those encountered with attach (variables and data frames being used that you were not aware had been loaded).

To solve this problem, the easiest option is to clear the workspace (see also Chapter 1) with:

```
> rm(list = ls(all = TRUE))
```

Now quit R and save the (empty) workspace. The alternative is to locate the .RData file and manually delete it from Windows Explorer. In our computer (using VISTA), it would be located in the directory: C:/Users/UserName. Network computers and computers with XP are likely to have different settings for saving user information. The best practice is simply to avoid saving the workspace.

References

Barbraud C, Weimerskirch H (2006) Antarctic birds breed later in response to climate change. *Proceedings of the National Academy of Sciences of the USA* 103: 6048–6051.

Bivand RS, Pebesma EJ, Gómez-Rubio V (2008) *Applied Spatial Data Analysis with R*. Springer, New York.

Braun J, Murdoch DJ (2007) *A First Course in Statistical Programming with R*. Cambridge University Press, Cambridge.

Chambers JM, Hastie TJ (1992) *Statistical Models in S*. Wadsworth & Brooks/Cole Computer Science Series. Chapman and Hall, New York.

Claude J (2008) *Morphometrics with R*. Springer, New York.

Cleveland WS (1993) *Visualizing Data*, Hobart Press, Summit, NJ, 360 pp.

Crawley MJ (2002) *Statistical Computing. An Introduction to Data Analysis Using S-Plus*. Wiley, New York.

Crawley MJ (2005) *Statistics. An Introduction Using R*. Wiley, New York.

Crawley MJ (2007) *The R Book*. John Wiley & Sons, Ltd., Chichester.

Cruikshanks R, Laursiden R, Harrison A, Hartl MGH, Kelly-Quinn M, Giller PS, O'Halloran J (2006) *Evaluation of the use of the Sodium Dominance Index as a Potential Measure of Acid Sensitivity (2000-LS-3.2.1-M2) Synthesis Report*, Environmental Protection Agency, Dublin, 26 pp.

Dalgaard P (2002) *Introductory Statistics with R*. Springer, New York.

Everitt BS (2005) *An R and S-Plus Companion to Multivariate Analysis*. Springer, London.

Everitt B, Hothorn T (2006) *A Handbook of Statistical Analyses Using R*. Chapman & Hall/CRC, Boca Raton, FL.

Faraway JJ (2005) *Linear Models with R*. Chapman & Hall/CRC, FL, p 225.

Fox J (2002) *An R and S-Plus Companion to Applied Regression*. Sage Publications, Thousand Oaks, CA.

Gentleman R, Carey V, Huber W, Irizarry R, Dudoit S, editors (2005) *Bioinformatics and Computational Biology Solutions Using R and Bioconductor*. Statistics for Biology and Health. Springer-Verlag, New York.

Gillibrand EJV, Bagley P, Jamieson A, Herring PJ, Partridge JC, Collins MA, Milne R, Priede IG (2006) Deep Sea Benthic Bioluminescence at Artificial Food falls, 1000 to 4800 m depth, in the Porcupine Seabight and Abyssal Plain, North East Atlantic Ocean. *Marine Biology* 149: doi: 10.1007/s00227-006-0407-0

Hastie T, Tibshirani R (1990) *Generalized Additive Models*. Chapman and Hall, London.

Hemmingsen W, Jansen PA, MacKenzie K (2005) Crabs, leeches and trypanosomes: An unholy trinity? Marine Pollution Bulletin 50(3): 336–339.

Hornik K (2008) The R FAQ, http://CRAN.R-project.org/doc/FAQ/

Jacoby WG (2006) The dot plot: A graphical display for labeled quantitative values. *The Political Methodologist* 14(1): 6–14.

Jolliffe IT (2002) *Principal Component Analysis*. Springer, New York.

A.F. Zuur et al., *A Beginner's Guide to R*, Use R,   207
DOI 10.1007/978-0-387-93837-0_BM2, © Springer Science+Business Media, LLC 2009

Keele L (2008) *Semiparametric Regression for the Social Sciences*. Wiley, Chichester, UK.

Legendre P, Legendre L (1998) *Numerical Ecology* (2nd English edn). Elsevier, Amsterdam, The Netherlands, 853 pp.

Lemon J, Bolker B, Oom S, Klein E, Rowlingson B, Wickham H, Tyagi A, Eterradossi O, Grothendieck G, Toews M, Kane J, Cheetham M, Turner R, Witthoft C, Stander J, Petzoldt T (2008) Plotrix: Various plotting functions. R package version 2.5.

Loyn RH (1987) Effects of patch area and habitat on bird abundances, species numbers and tree health in fragmented Victorian forests. In: Saunders DA, Arnold GW, Burbidge AA, Hopkins AJM (eds) *Nature Conservation: The Role of Remnants of Native Vegetation*. Surrey Beatty & Sons, Chipping Norton, NSW, pp. 65–77.

Magurran, AE (2004) *Measuring Biological Diversity*. Blackwell Publishing, Oxford, UK.

Maindonald J, Braun J (2003) *Data Analysis and Graphics Using R* (2nd edn, 2007). Cambridge University Press, Cambridge.

Mendes S, Newton J, Reid R, Zuur A, Pierce G (2007) Teeth reveal sperm whale ontogenetic movements and trophic ecology through the profiling of stable isotopes of carbon and nitrogen. Oecologia 151: 605–615.

Murrell P (2006) *R Graphics*. Chapman & Hall/CRC, Boca Raton, FL.

Nason GP (2008) *Wavelet Methods in Statistics with R*. Springer, New York.

Oksanen J, Kindt R, Legendre P, O'Hara B, Simpson GL, Solymos P, Stevens MHH, Wagner H (2008) Vegan: Community Ecology Package. R package version 1.15-0. http://cran.r-project.org/, http://vegan.r-forge.r-project.org/

Originally Michael Lapsley and from Oct 2002, Ripley BD (2008) RODBC: ODBC Database Access. R package version 1.2-4.

Pinheiro J, Bates D, DebRoy S, Sarkar D and the R Core Team (2008) nlme: Linear and nonlinear mixed effects models. R package version 3.1-88.

R-Core Members, Saikat DebRoy, Roger Bivand and Others: See Copyrights File in the Sources (2008) Foreign: Read Data Stored by Minitab, S, SAS, SPSS, Stata, Systat, dBase, ... R package version 0.8-25.

Lemon J, Bolker B, Oom S, Klein E, Rowlingson B, Wickham H, Tyagi A, Eterradossi O, Grothendieck G, Toews M, Kane J, Cheetham M, Turner R, Witthoft C, Stander J and Petzoldt T (2008). plotrix: Various plotting functions. R package version 2.5.

Pinheiro J, Bates D (2000) *Mixed Effects Models in S and S-Plus*. Springer-Verlag, New York, USA.

Quinn GP, Keough MJ (2002) *Experimental Design and Data Analysis for Biologists*. Cambridge University Press, Cambridge.

R Development Core Team (2008) *R: A Language and Environment for Statistical Computing*. R Foundation for Statistical Computing, Vienna, Austria. ISBN 3-900051-07-0, URL http://www.R-project.org

Reed JM, Elphick CS, Zuur AF, Ieno EN, Smith GM (2007) Time series analysis of Hawaiian waterbirds. In: Zuur AF, Ieno EN, Smith GM (eds) *Analysing Ecological Data GM*. Springer, New York.

Roulin A, Bersier LF (2007) Nestling barn owls beg more intensely in the presence of their mother than their father. Animal Behaviour 74: 1099–1106.

Sarkar D (2008) *Lattice: Lattice Graphics*. R package version 0.17-2

Shumway RH, Stoffer DS (2006) *Time Series Analysis and Its Applications with R Examples*. Springer, New York.

Sikkink PG, Zuur AF, Ieno EN, Smith GM (2007) Monitoring for change: Using generalised least squares, non-metric multidimensional scaling, and the Mantel test on western Montana grasslands. In: Zuur AF, Ieno EN, Smith GM (eds) *Analysing Ecological Data GM*. Springer, New York.

Spector P (2008) *Data Manipulation with R*. Springer, New York.

Venables WN, Ripley BD (2002) *Modern Applied Statistics with S* (4th edn). Springer, New York. ISBN 0-387-95457-0

Verzani J (2005) *Using R for Introductory Statistics*. CRC Press, Boca Raton.

Vicente J, Höfle U, Garrido JM, Fernández-de-Mera IG, Juste R, Barralb M, Gortazar C (2006) Wild boar and red deer display high prevalences of tuberculosis-like lesions in Spain. Veterinary Research 37: 107–119.

Wood SN (2006) *Generalized Additive Models: An Introduction with R*. Chapman and Hall/ CRC, NC.

Zar JH (1999) *Biostatistical Analysis* (4th edn). Prentice-Hall, Upper Saddle River, USA.

Zuur AF, Ieno EN, Smith GM (2007) *Analysing Ecological Data*. Springer, New York, 680p.

Zuur AF, Ieno EN, Walker NJ, Saveliev AA, Smith G (2009) *Mixed Effects Models and Extensions in Ecology with R*. Springer, New York.

Index

Note: Entries in **bold** refer to command/function/argument.

Bayesian Computation with R, Second Edition

Jim Albert

This book is a suitable companion book for an introductory course on Bayesian methods and is valuable to the statistical practitioner who wishes to learn more about the R language and Bayesian methodology. The second edition contains several new topics such as the use of mixtures of conjugate priors and the use of Zellner's g priors to choose between models in linear regression. There are more illustrations of the construction of informative prior distributions, such as the use of conditional means priors and multivariate normal priors in binary regressions. The new edition contains changes in the R code illustrations according to the latest edition of the LearnBayes package.

2009. Second ed. Approx. 308 p. (Use R) Softcover
ISBN 978-0-387-92297-3

Mixed Effects Models and Extensions in Ecology with R

Alain F. Zuur, Elena N. Ieno, Neil J. Walker
Anatoly A. Saveliev, and Graham M. Smith

The first part of the book is a largely non-mathematical introduction to linear mixed effects modelling, GLM and GAM, zero inflated models, GEE, GLMM and GAMM. The second part provides ten case studies that range from koalas to deep sea research. These chapters provide an invaluable insight into analysing complex ecological datasets, including comparisons of different approaches to the same problem. By matching ecological questions and data structure to a case study, these chapters provide an excellent starting point to analysing your own data.

2009. Approx 530 p., Hardcover
ISBN 978-0-387-87457-9

R Through Excel
A Spreadsheet Interface for Statistics, Data Analysis, and Graphics

Richard M. Heiberger
Erich Neuwirth

In this book, the authors build on RExcel, a free add-in for Excel that can be downloaded from the R distribution network. RExcel seamlessly integrates the entire set of R's statistical and graphical methods into Excel, allowing students to focus on statistical methods and concepts and minimizing the distraction of learning a new programming language. The book is designed as a computational supplement to introductory statistics texts and the authors provide RExcel examples covering the topics of the introductory course.

2009. XXIV, 338 p. (Use R) Softcover
ISBN: 978-1-4419-0051-7

springer.com

Applied Econometrics with R

Christian Kleiber
Achim Zeileis

This book covers a variety of regression models (beginning with the classical linear regression model estimated by ordinary least quares,) regression diagnostics and robustness issues, the nonlinear models of microeconomics (Logit, Probit, Tobit, and further models), time series and time series econometrics (including unit roots and cointegration).

Content: Introduction.- Basics.- Linear regression.- Diagnostics and alternative methods of regression.- Models of microeconometrics.- Time series.- Programming your own analysis.- References.- Index.

2008.X, 222 p. (Use R!) Softcover
ISBN: 978-0-387-77316-2

Finite Mixture and Markov Switching Models

Sylvia Frühwirth-Schnatter
WINNER OF THE 2007 DEGROOT PRIZE!

Content: Finite Mixture Modelling.- Statistical Inference for a Finite Mixture Model with Known Number of Components.- Practical Bayesian Inference for A Finite Mixture Model With Known Number of Components.- Statistical Inference for Finite Mixture Models Under Model Specification Uncertainty.- Computational Tools for Bayesian Inference for Finite Mixture Models Under Model Specification Uncertainty.- Finite Mixture Models With Normal Components.- Data Analysis Based on Finite Mixtures.- Finite Mixtures of Regression Models.- Finite Mixture Models with Non-Normal Components.- Finite Markov Mixture Modelling.- Statistical Inference for Markov Switching Models.- Non-Linear Time Series Analysis Based on Markov Switching Models.- Switching State Space Models.

2006. XX, 492 p. (Springer Series in Statistics) Hardcover
ISBN: 978-0-387-32909-3

R Through Excel

Richard M. Heiberger
Erich Neuwirth

The book is designed as a computational supplement to introductory statistics texts and the authors provide RExcel examples covering the topics of the introductory course.

Content: Getting started.- Using RExcel and R Commander.- Getting data into R.- Normal and t Distributions.- Normal and t workbook.- t-tests.- One-way ANOVA.- Simple linear regression.- What is least squares?- Multiple regression-Two X-variables.- Polynomial regression.- Multiple regression-Three or more X-variables.- Contingency tables and the Chi-Square test.- Appendix: Installation of RExcel.

2009. XXIV, 338 p. Softcover (Use R)
ISBN: 978-1-4419-0051-7

Lightning Source UK Ltd.
Milton Keynes UK
UKHW02f1547110618
323801UK00004B/101/P